THE TOMB OF THE MILI MONGGA

THE TOMB
OF THE
MILI MONGGA

*Fossils, Folklore, and Adventures
at the Edge of Reality*

Samuel Turvey

BLOOMSBURY SIGMA
LONDON · OXFORD · NEW YORK · NEW DELHI · SYDNEY

BLOOMSBURY SIGMA
Bloomsbury Publishing Plc
50 Bedford Square, London, WC1B 3DP, UK
29 Earlsfort Terrace, Dublin 2, Ireland

BLOOMSBURY, BLOOMSBURY SIGMA and the Bloomsbury Sigma logo are
trademarks of Bloomsbury Publishing Plc

First published in the United Kingdom in 2024

A catalogue record for this book is available from the British Library

Library of Congress Cataloguing-in-Publication data has been applied for

ISBN: HB: 978-1-399-40977-3; ePub: 978-1-399-40979-7; ePDF: 978-1-399-40974-2;
audio: 978-1-399-40976-6

2 4 6 8 10 9 7 5 3 1

Typeset by Deanta Global Publishing Services, Chennai, India
Printed and bound in Great Britain by CPI Group (UK) Ltd, Croydon CR0 4YY

Bloomsbury Sigma, Book Seventy-Nine

To find out more about our authors and books visit www.bloomsbury.com
and sign up for our newsletters

For Umbu

And for Ben Collen and Georgina Mace,
who I hope would have chuckled if they knew I was
writing this book.

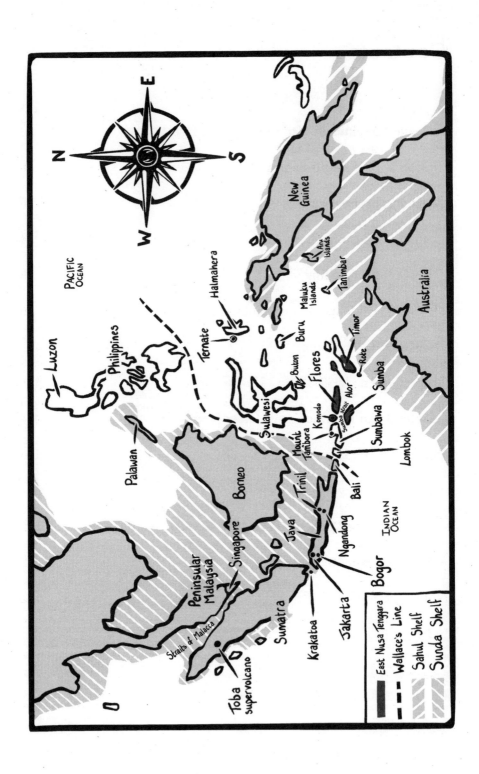

Contents

Prologue: Anselm and Gaunilo   10

Chapter 1: Splendid Isolation   18

Chapter 2: Sumba, East of Java   36

Chapter 3: Glutton-Granny   54

Chapter 4: Storytelling   78

Chapter 5: Rodents of Unusual Size   96

Chapter 6: Tulang Junkie   122

Chapter 7: The Wall of the Mili Mongga   144

Chapter 8: An Interlude with Giant Rats   174

Chapter 9: The Island of the Day Before   194

Chapter 10: They Might Be Giants   234

Chapter 11: The Perfect Island – A Fairy Tale for Biologists   256

Acknowledgements   269

Notes   272

Index   294

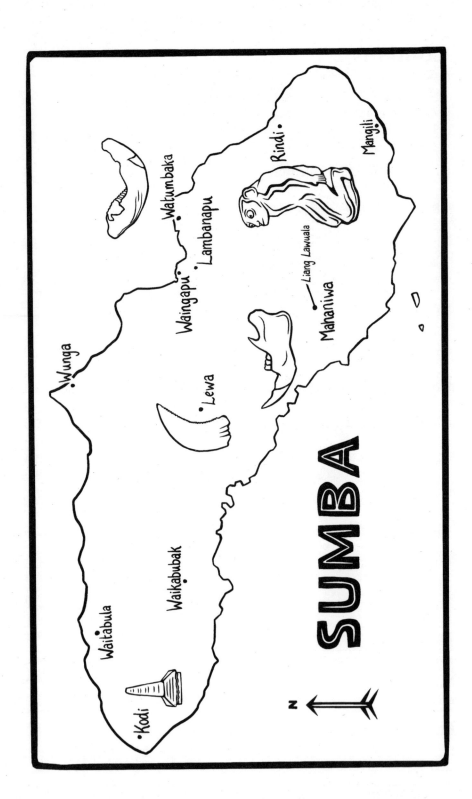

SUMBA

Waitabula
Kodi
Waikabubak
Wunga
Lewa
Waingapu
Lambanapu
Watumbaka
Rindi
Mangili
Liang Lawuala
Mahaniwa

N

Sir, I invite your highness and your train
To my poor cell, where you shall take your rest
For this one night; which, part of it, I'll waste
With such discourse as, I not doubt, shall make it
Go quick away – the story of my life,
And the particular accidents gone by
Since I came to this isle.

William Shakespeare, *The Tempest*

Well, that's how history's written, isn't it?

Marcel Proust, *In the Shadow of Young Girls in Flower*
(after a quote by Voltaire)

Anselm and Gaunilo

The universe is full of magical things patiently waiting for our senses to grow sharper.

Eden Phillpotts, *A Shadow Passes*[*]

The only part of a story that is true is the part the listener believes.

Hermann Hesse

Matius and Yakobus were back from church and already waiting when our car pulled up in the village. They opened the door and squeezed in, somehow making room on the back seat of the already-packed vehicle. Resting his parang between his knees so that the blade pointed downwards, Matius smiled broadly at us, his teeth stained a deep red. Following directions from the two men, we turned down a dirt track at the edge of the village, stopping briefly for Matius to jump out and run up to a shack with a corrugated metal roof to buy cigarettes and *sirih pinang* for the day's hike. Bob Marley blared from the car stereo as we bounced along the dusty track, until Matius called out for us to stop as we neared another hut with thin walls made of plaited cane. High above, a sea eagle circled over the dry landscape.

We set off on foot along the valley behind the hut, walking past an abandoned rice paddy and mounds of dry goat dung. As I pushed my way through tall grass that reached to my shoulders, a pleasant minty smell rose up from the vegetation, and a huge, bright yellow butterfly as big as a bird beat its way lightly past me. After a while, when we

---

[*] This quote is usually attributed to W. B. Yeats, and is thought to describe Yeats' vision of a supernatural world rich with unknown possibilities on the edge of human perception. In fact it is by Eden Phillpotts, an English novelist from the early twentieth century, and refers to the power of progressive scientific investigation to discover what was previously unknown – a completely opposite intent. This confusion over both origin and meaning makes it an appropriate quote with which to start this book.

had crossed a small stream, the path entered the forest. Here the air was still, except for the hum of insects and an occasional faint wind, with the silence broken only by the rhythmic crunching of our feet on dead leaves. Sweat ran down my back. The path turned uphill and became indistinct, and we clambered onward over sharp rocks sticking up from the forest floor. The hill was increasingly steep, and the heavy sounds of my laboured breathing filled my ears. Suddenly, we had made it; the trees were behind us, and we were through to a grassy clearing on the ridge of the hill. Looking back the way we had come, over the dry forest now spread out below us, the sun caught on the tiny steep metal roofs of the village in the distance. I pulled the dusty plastic bottle from the side pocket of my rucksack and drank down a long, refreshing draught of lukewarm water. Ahead, the dry hills stretched on to the horizon, with scrappy patches of forest clinging to the valleys that dropped down from either side of the ridge. I offered my bottle to Umbu. He looked at me. 'I don't need water.' The serious expression on his face changed into a conspiratorial grin. 'I just need bones!'

The trail followed the ridge, heading towards a large solitary coconut tree in the far distance. The grass on the hilltop was sun-bleached, dead, white; the soil was grey and ashy. There was no colour anywhere, except for a pale yellow tinge to the plants growing on the lower slopes of the hill. A few clouds drifted far above us, disinterested. Bee-eaters perched in the tops of dead trees sticking out of the side of the ridge. We walked silently in single file past an old, dug-out grave. Heat tightened around my head like a band.

On the long drive back to Waingapu the previous evening, after we had spent the day talking to the villagers about the purpose of our visit, I sat in the car shelling peanuts contemplatively. The light drained from the landscape as the sun set, leaving the world outside the car flat and ghostly. Somehow, exploring an ancient abandoned graveyard felt worse on a Sunday. But the story that Matius and Yakobus had told about what they'd seen was too tantalising to ignore. Would this finally provide the answer to the mystery we'd heard so much about throughout our time on Sumba?

At the head of the line of walkers, Matius suddenly dipped off the ridge down the side of the hill on our right, leading us back into the forest. There was no path now, just snags and rocks and potholes, but after the heat of the hill it was refreshingly cool and dark under the trees. There were new sounds here, tinkling and bubbling above us, the birds making these calls themselves unseen. The forest floor was made of boulders of ancient coral rubble, attesting to the antiquity of the landscape we were hiking through and the changes it had experienced across gulfs of time too great to imagine. A new trail emerged between the rocks, and we pressed on under huge trees. I stepped over dry brush, ducking under branches and vines as thorns snagged at my arms. These were hornbill trees, Matius said; the forests behind the village still contained 10 pairs of the rare species of hornbill that was only found on Sumba, and they needed massive trees like these in which to nest and roost.

I clambered over dry brown lianas wrapped around a low overgrown stone wall, the entrance to the long-abandoned old village in the forest. Inside the wall, the floor was made up of broken slabs of ancient graves. Anxious bird calls, bouncing back and forth to each other in the trees, rose to a crescendo. Matius yelled out in amusement – a megapode scampered low across the trail in front of him, disappearing off to the right. Thin saplings grew up through the grave rubble. Some of the tombs were still upright and intact between the trees, like huge stone dinner tables on massive pillar legs, covered in dry curled leaves. A line of ants marched over the surface of the nearest grave. Ritual cupmarks were carved on the capstone where *sirih pinang* could be left for the dead to chew. A hornbill called from somewhere far off through the forest.

Under the cool trees, I walked past a jumble of broken slabs from damaged tombs. How old was this rubble of time? Suddenly there was a shout from up ahead. Matius placed the coils of rattan that he was collecting from the forest down on a capstone. Umbu ran over and stood next to Yakobus. 'Pak Sam!' Umbu called to me. 'The bones. They are here.'

In 1078, an abbot named Anselm from Bec, an eleventh-century centre of learning in northern France, put forward what is now considered to be western philosophy's first ontological argument – an argument that aims to prove the existence of God. Imagine the greatest being you possibly can, Anselm proposed. Even if you don't believe in God, such an entity can still exist in your mind. But if you can conceive of such an imaginary being, then an even greater being *must* be possible. Existence is perfection, Anselm argued; so something that actually exists in reality is inherently greater than an imaginary version that exists merely as an idea in the mind. Ergo, concluded Anselm, this is evidence for God.

Anselm's argument constituted robust and challenging logic in the eleventh century. He went on to become Archbishop of Canterbury and was then made a saint, and his ontological argument became the subject of fierce debate – did it *really* provide irrefutable proof of God's existence? Not everyone agreed, even at the time. Soon after Anselm laid out his theological proof, another religious thinker, a French Benedictine monk called Gaunilo, responded with a criticism called *In Behalf of the Fool*. Gaunilo considered that Anselm's logic leads to the conclusion that many things exist which certainly do not, and he picked apart his argument by parodying it with another thought experiment. Instead of the greatest being, try to imagine the most perfect island, challenged Gaunilo. If we can conceive of such an island paradise, it must – by Anselm's reasoning – therefore also exist in reality, as an existing island will by its very nature be more perfect than an imaginary one.

Did Gaunilo's counter-argument fatally undermine Anselm's ontological proof for the existence of God? Anselm himself didn't think so; the debate rumbled on through the centuries, and ultimately helped to shape the course of modern philosophical thinking. In retrospect though, this medieval disagreement around the thorny question of what constitutes valid evidence of God raises very different points of interest. This was an argument of paramount importance throughout the Middle Ages and beyond – but it feels incongruous, even alien, to the types of reasoning that we now use to interpret the world around us, and even to what constitutes a meaningful question for many of us. Who today would think to base a logical argument around the 'irrefutable' premise that existence constitutes perfection? Thinking about Anselm's supposed proof makes us aware of the extent to which our worldviews and perspectives differ from those of other peoples – both the societies that existed in the past, and also potentially the cultures that still exist in other parts of the world today. Are any of these modes of thought, these differing assumptions about reality, actually closer than any others to some sort of objective 'truth'? And can we ever even manage to comprehend and see beyond the paradigms of the specific time and place in which we find ourselves?

But Gaunilo's critique also raises the fascinating possibility of something that has been overlooked through the centuries by the various protagonists involved in this ontological debate. His thought experiment was intended to be absurd. But what if there *is* actually a perfect island out there somewhere? What might such a place be like, in imagination and in reality? And if you could find it, somewhere out there over the horizon, what would be waiting to greet you when you arrived? Might you end up finding evidence for God – or something even more strange? Or would its fantastic inhabitants turn out just to be imaginary after all?

To a biologist, all islands might be considered perfect. These isolated, fascinating places have set the scene for evolution to run wild in strange new directions and produce some of the world's most unexpected and remarkable biodiversity. Or at least these island ecosystems used to be perfect, until people came along.

This book is about my explorations of an island on the other side of the world, to try to understand what kinds of unique species used to inhabit its remote landscapes, and what happened to these now-vanished animals. But it isn't just a story about biology or biologists, even though I thought it would be when I started out on my adventure. There's plenty of science and natural history in the pages that follow, which can hopefully also serve to illustrate the steps through which knowledge accumulates and science progresses; how sources of inspiration might be unexpected, requiring new leads to be followed in unplanned directions when confronted with things that we can't easily rationalise or comprehend; and how possible explanations for our observations are often competing and hopelessly confusing. What are the different types of evidence we need to consider in order to reconstruct both the present and the past? And how can we understand the truth – indeed, what does this even mean, and is it ever really possible to achieve? What began as a study into biodiversity, fossils and extinction had to take on other intellectual frameworks and appreciate very different cultural perspectives. It also became a wider story about how people interpret the world around them – how they understand nature, ancient remains and traces from the past, other people and peoples, and even themselves – and how we can't presume that there is just one obvious, intuitive way of doing this. Just how subjective and relative are our shifting worldviews – between individuals, across cultures, and through history? And ultimately, what do you do when you learn about things that can't possibly be true?

In *The Adventure of the Sussex Vampire*, one of Arthur Conan Doyle's final Sherlock Holmes stories (and one with its own 'mixture of the modern and the mediæval, of the practical and of the wildly fanciful'), Holmes famously informed Watson about the case of the giant rat of Sumatra, 'a story for which the world is not yet prepared'. My story is about giant rats on another Indonesian island, along with lots of other things that might either be real or not, or maybe somewhere in-between, and for which I don't know if I was really prepared. You might even say, if you really wanted, that this is a story about stories – a true fairy tale. Although that might be taking it just a bit too far.

CHAPTER ONE

SPLENDID ISOLATION

Strangely, from your little island in space, you were gone forth into the dark, great realms of time, where all the souls that never die veer and swoop on their vast, strange errands. The little earthly island has dwindled, like a jumping-off place, into nothingness, for you have jumped off, you know not how, into the dark wide mystery of time, where the past is vastly alive, and the future is not separated off.

D. H. Lawrence, 'The Man Who Loved Islands'

Islands are powerful places. These little worlds, bound in and isolated by nature, and possessing an intangible, magical essence – where things are both familiar and somehow *different* – have provided fertile ground for the human imagination for as long as we have had poetry, literature and stories. An island, entire of itself, represents the world in miniature; these 'tiny pieces of land, the existence of which imagination can just about hang on to',[1] act as metaphors to provide new clarity about life when the details are simplified and stripped away. Free from the influence of the mainland, islands are stepping stones to other worlds. They inspire excitement and dreams of beguiling paradises, buried treasure and hidden secrets. They are lands of mysterious dangers and monsters. Their remoteness, isolation and simplicity have presented a primal challenge to marooned Robinson Crusoes and Swiss Family Robinsons, encouraging innovation and novel solutions to daily survival. They provide a fertile stage for experimentation, showing us alternate outward or inward realities and how these realities could be reached – from new models of political philosophy and how society might function to deeply personal revelation about identity, explored in solitude when the wider world becomes distant and its noise is quieted. Even the land underfoot is mutable and strange; it can be inundated by the waves or change unexpectedly into a kraken or aspidochelone, huge monsters that will drag the unwary down with them into the depths. We give islands our own meaning, with the specific flavour of different island narratives framed as fantasy, romance, allegory or satire, and addressing changing contemporary concerns of political expansion, exploration, colonialism or individualism. And from Gulliver to Pincher Martin, are the strange events experienced on remote islands even real at all?

The reasons that islands appeal so strongly to our imagination, and the ways in which they have been used in literature to explore differing aspects of the human condition, are mirrored in their importance for evolutionary science. For well over a century, islands have been recognised by scientists as being able to provide crucial insights into the processes by which new species are generated. Islands possess two key ecological characteristics that make this possible: isolation and simplicity.

All islands, by their very nature, are isolated in space by water barriers. This geographic isolation results in reproductive isolation: after a population of a particular species finds itself on an island, genes are rarely – if ever – exchanged through reproduction with individuals from the wider outside world, and genetic changes that might arise are not diluted out or modified. Instead, this genetic novelty may spread through the island population and become fixed in all of its descendants – either because it confers some specific environmental advantage, or simply as a result of mating within a small isolated population (especially if individuals are related). Populations of species that can move freely over both land and water, such as seabirds, are less affected by an island's isolation; but even for these species, site fidelity to traditional nesting grounds means that distinct genetic lineages associated with different islands can still often develop.

In addition to isolation in space, the accumulation of genetic differences in island populations is also dependent upon another type of isolation – isolation in time. The amount of this second type of isolation that a population will experience depends upon the geological origins of the island where it lives. Although all islands are superficially similar to one another – basically, they're all lumps of land surrounded by water – they fall into a series of distinct geological types associated with very different histories of isolation.

Islands with the shortest history of isolation are continental islands, which form part of the same continuous tectonic block of continental crust as nearby continental mainlands. These islands are only isolated by shallow seaways, which ebb and flow over time, and so periodically reconnect the islands to the mainland. Tidal islands, such as St Michael's Mount in Cornwall, are reconnected every day to the mainland by causeways that emerge above water at low tide. Other continental islands such as the British Isles have remained isolated for far longer than recorded human history, but their island identity is still transient in geological terms. They have been reconnected repeatedly to the neighbouring continental mainland during numerous intervals of sea-level change over the past couple of million years, as the Earth has moved through Ice Age glaciation cycles driven by gradual variations in its orbit around the sun, which caused huge amounts of water to be periodically frozen and then released from the ice caps. At the moment we are in a high sea-level 'warm phase' of this cycle – the current Holocene Epoch of geological time, which commenced 11,700 years ago, and represents the period of broadly modern-day environmental conditions. However, as recently as 20,000 years ago during the preceding Late Pleistocene Epoch, the world was a much colder, drier place, with global sea levels lowered by up to 130m; in place of today's wetter geography, continental islands were instead connected by exposed land bridges to the edges of the continents. The British Isles themselves were joined to what is now the Netherlands and Denmark by a low-lying landmass called Doggerland, which was only inundated by rising sea levels within the last 9,000 years. Such a recent history of isolation has left little time for populations of animals and plants found on continental islands to accumulate genetic novelties and become evolutionarily distinct from populations on neighbouring continents.

Other islands, such as Madagascar and the islands of New Zealand, are also composed of continental crust. However, these continental fragments or microcontinents have not been connected to continental plates for a very long time; they were ripped free by powerful tectonic forces during the distant prehistoric past and have followed adventurous geological journeys of their own ever since, propelled by continental drift to inch along in isolation over aeons of geological time. The last time that both Madagascar and New Zealand formed part of a larger continental landmass was when the supercontinent of Gondwana occupied much of the southern hemisphere and dinosaurs walked the

Earth, allowing ample time for some isolated populations to die out and others to evolve into new species. Many of today's vagrant microcontinents serve as the final refuges for evolutionarily ancient lineages that vanished long ago from other parts of the world – they have become museums of ancient biodiversity. For example, the tuatara of New Zealand superficially looks like a lizard (and was originally classified as such), but is in fact the only surviving member of an entire reptile order, the Rhynchocephalia, which first appeared around 200 million years ago and once contained a diversity of species found all over the world.

The third category of islands, oceanic islands, have never been part of a continent or continental shelf. Instead, they are formed by powerful underwater tectonic forces that push upwards and create new land in the middle of the ocean. Such islands are typically produced by volcanic activity, either along tectonic boundaries or at volcanic hotspots in the middle of tectonic plates. Some are several million years old, such as the north-western islands of the Hawaiian archipelago, while others have emerged from the waves within the past century. Because these islands have never been connected to a neighbouring landmass, the only way that animals or plants can reach them is via overwater dispersal. James Henry Trotter rode a giant peach to cross the Atlantic Ocean in Roald Dahl's story; zoological Trotters might instead get blown overwater by a hurricane (if they are birds or insects), or arrive on handy rafts or mats of vegetation that have drifted into the open ocean (if they are reptiles, mammals, or other species generally unable to fly). Seeds and other plant propagules could be carried to oceanic islands by favourable oceanic currents or in bird droppings.

Overwater colonisation on a convenient 'floating island' may seem highly implausible, and it's certainly a rare chance event. However, the older literature contains a surprisingly large number of reports of wild animals being accidentally transported from place to place in this way. Following the flood of the Mississippi River in 1874, observers in the *Mississippi Sound* noted that 'For miles were seen logs, driftwood and patches of turf and soil floating out into the gulf, filled with live animals, which clung to their frail barques with the tenacity of shipwrecked mariners. Among the animals were seen rats, raccoons, possums, rabbits, alligators, and moccasin snakes, in uncounted numbers.'[2] A ship travelling from Cuba to Philadelphia in 1902 encountered a floating island with many upright palm trees on it about 50km from the island of San Salvador; when some of the crew rowed over to investigate, they found that it was carrying numerous monkeys, some of which reportedly 'threw coconuts at them'.[3] In 1825, a floating raft of vegetation that was washed down the Paraná River to the Iglesia de San Francisco in Santa Fe was carrying a jaguar, which leapt ashore and killed two of the church's friars.[4] Most incredible of all, a floating island in Bangladesh that travelled down a river to Chittagong in 1868 was reportedly carrying ... a live Sumatran rhino.[5] The animal, a female named Begum, was dragged out of some quicksand near the mouth of the river[6] and sent to London Zoo in 1872, where she was described as a new species (now regarded as just a distinct northern subspecies of Sumatran rhino). She ended up living in the zoo until 1900, becoming the longest-lived captive rhino at the time.

There are even direct observations of successful overwater colonisations of new islands taking place in this manner. One of the best-known examples was witnessed in the Caribbean in 1995, when Hurricanes Luis and Marilyn swept past the island of Guadeloupe and uprooted a huge mat of trees and logs, which was carried more than 300km by powerful currents to the more northerly island of Anguilla. Clinging onto this raft of vegetation were at least 15 green iguanas, a species found on Guadeloupe but not Anguilla. Although they seemed weak and dehydrated, and some were injured, the iguanas were seen to climb ashore. The group of animals contained both males and females, and a population of iguanas descended from these founders still survives on Anguilla today.[7]

In practice, the differences between continental islands, microcontinents and oceanic islands aren't totally clear-cut. Overwater dispersal can contribute to the diversity of all island types. Some geologically ancient islands, such as Jamaica and New Caledonia, are thought to have been largely or completely inundated by the sea at some point in time after breaking free from their parent landmasses. Despite their different geological origins, such islands therefore act functionally like oceanic islands, becoming clean slates for new fauna and flora to colonise after their original diversity was drowned, although they sometimes still also retain hardy surviving relict lineages alongside new arrivals. And some islands are actually hybrids, with different parts of their landmasses having formed via different geological processes. For instance, around 41 million years ago, a tiny fragment of what had once been Gondwana broke away from the edge of a larger piece of this ancient supercontinent, which had itself already broken free and reached the western Pacific. This continental fragment drifted eastward and ploughed into the volcanic islands of the Tongan archipelago, which had formed through tectonic uplift along a plate boundary, where it became incorporated into the otherwise volcanic island of 'Eua. One of the reasons that 'Eua was discovered to have a different geological origin to the rest of the Tongan archipelago was because it is the only island in the group that contains native podocarps, a type of coniferous tree that evolved on Gondwana during the early Mesozoic Era long before the supercontinent broke apart.[8] These trees had unwittingly hitched a ride on the tiny scrap of continental crust as it drifted thousands of miles across the Pacific, to bear silent witness tens of millions of years later to the journey made by the bedrock beneath them.

Despite such complexities, this general framework of island types still helps us begin to understand how island animals and plants can be extremely distinctive and unusual, and how the level of differentiation they show depends upon the category of island they inhabit. To understand the generation of evolutionary novelty on islands more completely, we also need to consider the environmental conditions that exist on different islands.

As well as having been isolated for only a few thousand years, continental islands typically share very similar geologies, habitats and ecosystems to neighbouring continents. Essentially, they represent geographic extensions of continental environments, which just happen to exist on peninsulas that are periodically flooded. There is therefore rarely a strong evolutionary pressure for animals and plants on continental islands to adapt to unfamiliar environments, and they are very often unchanged in comparison to species on neighbouring continents. The British Isles have no endemic amphibian, reptile or mammal species (which evolved in situ and are found nowhere else), and only one endemic bird, the Scottish crossbill, which was until recently thought to be just a locally differentiated population rather than a fully distinct species. The rest of this island fauna is basically the same as that of northern France, with the exception of some species that have been historically eradicated in the British Isles (such as brown bears and wolves), and some that didn't quite manage to colonise after the end of the last Ice Age cold period before the English Channel flooded again (such as black woodpeckers and white-toothed shrews).

Vagrant microcontinents were also originally connected to prehistoric continents, and shared species in common many millions of years ago. However, these islands have been isolated for considerably longer, and have also moved long distances via tectonic action, often into regions with markedly different patterns of temperature, rainfall, ocean currents, and other key regulators of local climate. The tectonic processes that have shifted these islands around the globe have often also modified their physical structure and created novel environments, from new geological landscapes such as mountain ranges, to changes in composition of the rocks that form soils and support plants. Even if ancestral populations of continental animals and plants were carried along with such islands as they rifted away from other landmasses, they will have had to adapt dramatically to survive. Newly formed oceanic islands also present strikingly novel environments for any species that find themselves washed up or blown there, fuelling the evolution of new forms of life from such founders.

These abiotic factors play an important role in driving evolution on islands. However, island life is shaped even more distinctively by biotic factors – the diversity and identity of the different species that live alongside each other in island ecosystems. Ever since Charles Darwin used his metaphor of 'an entangled bank' in On the Origin of Species, competition for resources between individuals has been recognised as a crucial driver of evolutionary innovation through the mechanism of survival of the fittest. Continental ecosystems typically contain guilds composed of different species that all exploit similar resources in broadly similar ways – the diversity of large grazing mammals that coexist on an African savanna, or the different kinds of small seed-eating birds that visit your garden bird feeder. In contrast, island biotas are normally depauperate, containing only a subset of the animal and plant groups found on nearby continents, due to historical extinctions of former colonists and the restrictions posed by risky overwater dispersal. This means less competition for resources with members of other species, and correspondingly more competition with members of your own species. Furthermore, some species are worse at overwater colonisation than others. Many animal groups that are characteristic of continental landscapes have never managed to cross marine barriers and reach islands due to specific features of their ecology, leaving vacant niches in island ecosystems. For instance, most amphibians cannot tolerate saltwater, and some species that do occur on non-continental islands, such as the endemic frogs of Fiji, are much less dependent than other amphibians on fresh water (a scarce resource on many islands) – they lay their eggs in trees and lack a free-swimming tadpole stage. Some of the worst overwater colonisers are mammalian carnivores, which are found naturally on almost no continental fragments or oceanic islands (Madagascar is an important exception, where they have evolved into an endemic family of sometimes-giant mongooses).

Island-dwelling species show distinctive evolutionary patterns that reflect these different biotic and abiotic pressures. Successful colonists will regularly adapt to occupy unfilled niches, and will often independently evolve similar traits to those seen in unrelated continental species with comparable lifestyles – a process known as convergence. This morphological adaptation can be extreme, in both plants and animals. For instance, whereas conifers in most parts of the world are all broadly similar types of trees, bizarre aquatic conifers (*Retrophyllum minus*) and even shrubby parasitic conifers without any roots (*Parasitaxus usta*) have evolved on the remarkable island of New Caledonia, the ancient fragment of Gondwana that split off into the South Pacific.

Some of the weirdest examples of convergent evolution in island animals are seen on Madagascar. For example, the four genera of now-extinct palaeopropithecid lemurs, or 'sloth lemurs', evolved to hang beneath tree branches in this otherwise sloth-free island. These animals were so similar in their arboreal adaptations to the true sloths, the living *Bradypus* and *Choloepus* of Central and South America, that their postcranial remains were initially mistaken for the bones of sloths rather than primates.[9] However, whereas sloths possess long curved claws for gripping, sloth lemurs retained the tiny fingernails and toenails typical of other primates, and instead their hand and foot bones became extremely elongated to curve round branches. Madagascar also lacks woodpeckers, and the island's grub-hunting niche is instead filled in a very different way by another extremely unusual lemur, the aye-aye, which has evolved a remarkable toolkit for this purpose. It has a thin fourth finger for percussive foraging, to tap on trees to find grubs under bark, and enormous ears to listen for the faint chewing sounds made by its prey; it then gnaws a hole using its remarkable continually-growing 'buck-teeth' incisors; and probes for grubs in the newly-made hole using its massively elongated third finger, which rotates on a ball-and-socket joint and bears a hook-like fingernail.* It has recently been discovered that one of the aye-aye's wrist bones has also evolved into a 'pseudothumb' to help grip onto branches, to compensate for this extreme specialisation of its true fingers.[11] In a similar case of mistaken identity to the sloth lemurs, the aye-aye is so weird that it was originally thought to be a rodent, not a primate. In a remarkable further convergence, analogous adaptations for hunting wood-boring grubs are also seen in completely different mammals on another woodpecker-free island on the other side of the world: the trioks of New Guinea. Trioks are arboreal marsupials, members of the major mammal group that typically carry their young in a pouch, and are only distantly related to the aye-aye and other placenta-bearing mammals. (Marsupials and placentals, the two main mammalian radiations that exist today, are characterised by markedly different modes of reproduction and probably diverged about 150 million years ago.) However, trioks have independently evolved large ears, chisel-like gnawing teeth, and a hugely elongated finger to fish for grubs under bark, incredibly similar to the distinctive suite of characteristics seen in aye-ayes. To emphasise the convergence rather than common ancestry of this remarkable shared adaptation,

---

* It has recently been discovered that aye-ayes pick their noses with their third finger and then lick the collected nasal mucus, with the elongated finger able to reach almost to the back of the animal's throat, a behaviour formally described as 'rhinotillexis followed by mucophagy'.[10]

the triok's long grub-poking finger is the fourth finger rather than the third.

Evolutionary defences for avoiding predation, such as large wings and powerful flight muscles, are costly to maintain in terms of resources but are no longer needed in the absence of carnivores. Many island birds have therefore largely or completely lost the ability to fly, and other anti-predator adaptations have also disappeared in some island species. A striking example is provided by the extinct 'cave goat' *Myotragus* from the Balearic Islands of Majorca and Menorca. Whereas all continental ungulates (large hoofed herbivorous mammals) have eyes on the sides of their heads, allowing them to see across a very wide area when looking out for predators, *Myotragus* instead evolved forward-pointing eyes and stereoscopic vision – making it look uncannily like a person wearing a goat mask.[12] Increased competition with members of their own species has also led many island life-forms to invest more resources into producing fewer offspring, to give the next generation a better competitive edge; and in the absence of mammal predators, these offspring can grow slowly without fear of being picked off during vulnerable early life stages. The leg bones of flightless moa, giant birds that existed in New Zealand until relatively recently, contain growth rings that reveal they took up to a decade to become fully grown – whereas ostriches, the largest continental birds, become skeletally mature within 12 months.[13]

One of the most striking changes seen in island species is a change in body size. This occurs so regularly and predictably that it is often referred to as the 'island rule' or 'island law', formally known as Foster's rule. In an environment dominated by competition with other members of your own species, getting bigger can enhance your competitive edge. Because many characteristic large-bodied animals such as ungulates (other than *Myotragus*) typically fail to colonise islands, vacant niches and new resources are also available to exploit if you're big enough to reach them. Small animals therefore often increase in size on islands. Giant rats, giant dormice and other large rodents are characteristic components of island faunas, and the extinct hedgehog *Deinogalerix*, which lived 7–10 million years ago on the palaeo-island of Gargano (now part of southern Italy), grew to over

100 times the size of its inferred mainland ancestor.[14] Plants show comparable evolutionary patterns; Darwin marvelled at tree-sized daisies and sunflowers on the Galápagos Islands, fuchsias reach 15m in New Zealand, and the giant ginger grows to more than 8m tall in Fiji.

However, island lifeforms cannot keep getting bigger indefinitely. King Kong is very much a fictional species of island primate, not only because an ape of that size would face impossibly prohibitive mechanical and energetic constraints in order to exist at all, but also because island evolution is shaped by a further major limiting factor. Islands by their very nature are much smaller landmasses than continents, and so have proportionately fewer resources available to support animal and plant life. Island species have adapted to this limitation in resourceful ways. This was taken to one extreme by the weird cave goat *Myotragus*, which became cold-blooded to conserve energy as well as having forward-pointing eyes, meaning that it could probably only move around slowly like a lizard and was unable to run or jump.[15] A more widespread solution to resource limitation is another trend in body size seen across many island species: in direct contrast to the widespread pattern of small species increasing in size, large species typically shrink in size, in order to support viable populations of smaller individuals that require fewer resources to survive. This evolutionary response is facilitated by the general lack of native mammalian predators on islands, as there is no pressure to remain large to avoid being eaten. Many island species will therefore often converge, Goldilocks-like, upon a happy middle-size ground, neither too small nor too large but 'just right'.

Island dwarfing is illustrated in spectacular fashion in the recent fossil record by elephants and hippos. These large mammals are good swimmers and so managed to colonise numerous islands, where they shrank drastically over relatively rapid intervals of evolutionary time. (Island animals that have become evolutionarily 'shrunk' from larger-bodied ancestors are variously referred to as either dwarf or pygmy forms. Strictly speaking, these two terms have different anatomical meanings; I will refer to these animals as being dwarfed for consistency.) Fossils of tiny hippos are known from Madagascar, Cyprus, Crete, Sicily and Malta – where they were little bigger than the giant swans that also evolved in this upside-down ecosystem. Tiny proboscideans (elephants and their relatives) are known widely from the recent fossil record of islands in the Mediterranean, Southeast Asia, Siberia, California, and the north Pacific. The dwarf elephant *Palaeoloxodon falconeri*, also known from Malta and Sicily, measured a metre or less at shoulder height and was only 5 per cent the mass of its probable mainland ancestor, the straight-tusked elephant *Palaeoloxodon antiquus*.[16] Similar evolutionary patterns are seen even further back in the fossil record, with the remains of dwarfed dinosaurs (sauropods, hadrosaurs and other species) known from Romania's Hațeg Basin, which had been part of an island archipelago in the Tethys Sea at the end of the Cretaceous Period 70 million years ago. In fact, there is a close relationship between the total land area of an island and the body size of the largest animals it can support. The biggest terrestrial vertebrates known from Cuba and Hispaniola, the two largest Caribbean islands, were now-extinct dwarf ground sloths that weighed around 100kg, whereas the Balearic Islands supported nothing bigger than *Myotragus*, and the Galápagos and smaller Indian Ocean archipelagos were dominated by giant tortoises.[17] Based on the size of Skull Island shown on the map that's briefly on-screen in the original 1933 film, this island–size relationship suggests that King Kong wouldn't have reached my shoulder if he'd really existed, and would therefore have created substantially fewer problems for Fay Wray.

Island ecosystems have thus been shaped by very different forces compared to the continental systems with which we are more familiar; evolution has followed unusual paths in response to novel challenges, producing unexpected forms of biodiversity. Life does spectacular, ridiculous, experimental things on islands, making them endlessly fascinating to students of evolution. It's no surprise that H. G. Wells used an island setting for his imaginative Victorian evolutionary fable *The Island of Doctor Moreau*,* and islands have prompted questions about the origins of species for many hundreds of years, since long before Darwin visited the Galápagos. Peter Mundy, one of the first Europeans to taste tea or chocolate, visited remote Ascension Island in the south Atlantic in 1656 and ate a flightless rail (a small, hen-like wading bird)

---

* See also Karel Čapek's famous 1936 science-fiction parable *War with the Newts*, where the eponymous protagonists originated on an island off Sumatra.

he encountered there – which he reported was 'more than ordinary dainty meat, relishing like a roasting pig'[18] – but which also got him thinking about how these unusual birds had come about. 'The question is, how they should be generated, whether created there from the beginning, or that the earth produceth them of their own accord, as mice, serpents, flies, worms, insects, or whether the nature of the earth and climate have altered the shape and nature of some other fowl into this, I leave it to the learned to dispute of.'[19] The learned did indeed dispute; the Ascension rail is now named *Mundia* after its first recorded human predator, and islands have become recognised as fundamentally important natural laboratories, that can help us understand how biodiversity evolves in response to different conditions.

I have worked on islands for all of my career as a biologist. I did my first postdoctoral fellowship in New Zealand, and I've carried out long-term research on Caribbean islands and on Hainan, China's southernmost province, an island the size of Belgium in the South China Sea. However, the main focus of my work hasn't been to understand evolution, but instead its flipside: extinction. As well as displaying remarkable evolutionary patterns, island biodiversity is fragile; island species are tragically vulnerable to human actions. These two characteristics are not unrelated. Adaptations that were promoted on islands in the absence of predators, which was the evolutionary status quo for millions of years, suddenly became worse than useless when people and their animal companions turned up. Investing in only a few slow-growing offspring might be a good strategy for outcompeting other members of your own species, but it means that your young remain helpless for much too long when predators are around, and your population won't be able to bounce back quickly to replenish losses, so offtake quickly becomes unsustainable. And flightless ground-nesting birds or other ponderous island species have little chance against hungry sailors, feral dogs and cats, or rats that hitched a lift on boats and have a taste for eggs. For instance, sailors visiting the Galápagos Islands used to refer to the archipelago's giant tortoises as 'Galápagos mutton', hunting them in huge quantities for fresh meat, to obtain water from 'storage bags' in their necks and the dilute urine in their enormous bladders, and as provisions for ongoing sea voyages. Large numbers of tortoises were even transported to feed prospectors in California during the Gold Rush, and Darwin himself contributed to the slaughter (over 30 tortoises were collected for food by sailors on the HMS *Beagle*, although Darwin himself thought that the meat was generally 'indifferent').[20]

As well as their distinctive morphological and reproductive adaptations, island animals typically even lose their ability to react appropriately when predators suddenly appear – they are behaviourally naïve. This is poignantly illustrated in an account by Percy Lowe, an ornithologist who became Curator of Birds at London's Natural History Museum. In 1908, Lowe captured two hutias – large, guinea pig-like rodents only found in the Caribbean – from a tiny 2km-long limestone speck covered in thorn scrub and cacti called Little Swan Island, where they had been discovered 21 years earlier. One of the hutias died on the voyage back to England, but the other survived, and for some reason was considered sufficiently remarkable that it was exhibited before King Edward VII, who was accompanied by his favourite dog, Caesar.* Lowe wrote, with a combination of deference and understatement, that 'The rat exhibited not the faintest signs of fear or suspicion in the presence of the dog, or even of awe in the presence of His Majesty; and it seemed interesting to reflect, that if we exclude ourselves, "Caesar" was the very first mammal, of any kind other than his own race, that the rat had ever set eyes upon. While the dog was carefully held back, the rat came to the edge of the low table it was on, and putting its muzzle within an inch or two of the dog's, quietly inspected, what to it must have been a most extraordinary and surprising apparition.'[21] Lowe speculated that 'It is just possible … that Little Swan Island will, in the future, represent the last stronghold of this peculiar and old time race of rats, for here they are left absolutely unmolested; and no enemies, human or otherwise, seem likely to

---

* Caesar outlived Edward VII, and was so beloved that he led the late king's funeral procession ahead of the attending human royals and heads of state, which led to strong disapproval from Kaiser Wilhelm II; it has even been suggested that this perceived insult fuelled the Kaiser's hostility towards Britain and ultimately contributed to the outbreak of World War I. The dog was so famous that he was the subject of postcards, paintings, and even a handcrafted miniature Fabergé replica made of gold, enamel and gemstones. He now has his own Wikipedia page.

disturb them.'[22] Lowe's optimism was sadly misplaced. The lack of fear or suspicion exhibited by Little Swan Island hutias in the face of mammalian predators (and monarchs) proved rapidly to be their undoing. A box of unwanted cats was dropped off on Little Swan in the late 1950s, and hutias have never been seen there since.

The pitiful story of the Little Swan Island hutia is depressingly far from unique. The Stephens Island wren, a tiny flightless songbird, is famous for having supposedly been both discovered and wiped out by a single cat called Tibbles in 1894. Whilst this story is apocryphal, the truth is only marginally less simple and poignant. A pet cat (probably not called Tibbles) started bringing dead, sometimes half-eaten specimens of these weird mouse-like birds to its owner, the assistant lighthouse keeper David Lyall, who notified ornithologists about this remarkable find – but too late. By 1895, feral cats were running wild across the island and no live wrens could be found. More recently, and on the other side of the world, a unique subspecies of deer mouse used to be found on Estanque Island in the Gulf of California (contrary to what its name suggests, this was not an island mouse the size of a deer, but was just the size of an ordinary mouse). Until 1995, the mice were abundant and very tame, even feeding from the hands of visiting researchers. No scientists visited Estanque for a couple of years, but when they returned in 1998, they found no deer mice on the island; instead, they found a single female cat, probably released on the island by local fishermen, and cat scats full of mouse bones. The cat was removed in 1999, but by then her scats contained only lizard bones. The deer mice were extinct.[23]

On larger islands, extinctions might take more than just a single cat (or even a boxful), and can take years, decades, or even longer to play out. But the wider pattern is the same. Sadly, island animals are badly adapted for the consumption and competition of the human-dominated world. This is why many of the most remarkable examples of island evolution I've mentioned – Peter Mundy's tasty flightless rail, the sloth lemurs, the weird cave goat *Myotragus*, the moa, and the dwarf elephants, hippos and ground sloths – are now all gone forever. As well as the Little Swan Island hutia, about 100 other mammal species and 70 bird species are known to have become extinct in just the islands of the Caribbean since people arrived there a few thousand years ago. It has been suggested that possibly as many as 2,000 flightless rails alone might have disappeared following the overwater migration of Polynesian settlers as they spread across the many islands of the tropical Pacific (although this staggeringly high estimate has been challenged).[24] The catastrophic loss of island biodiversity – which included many remarkable ancient evolutionary pedigrees and many remarkable recent evolutionary radiations – has drastically pruned the global tree of life, and means that understanding the full scope of island evolution now requires a journey back in time: an exploration of the fossil record. And if you're interested in averting further extinctions, the last surviving fragments of incredible island biodiversity need your immediate attention.

This way of thinking about geographic patterns of conservation urgency also highlights a particular part of the world, as well as a particular type of ecosystem, which is in desperate straits. Species are, tragically, threatened all over the world by human activities. However, researchers who map the distributions of threatened species consistently identify Southeast Asia as an epicentre of extinction risk. This is a region of incredible biodiversity, where new discoveries are constantly being made. More than 2,200 new species were described here between 1997 and 2014, and huge numbers of unknown animals and plants undoubtedly still remain to be found in its tropical rainforests and coral reefs. However, Southeast Asia also contains the world's highest numbers of threatened mammal and plant species, and its unique biodiversity faces uniquely severe human pressures. More than half the world's human population, almost five billion people, is concentrated in Asia, posing incredible demands on space and resources. Destruction of the region's rainforests is ongoing, with deforestation rates among the highest in the world, and with global demand for timber, rubber and palm oil all major contributors to forest loss. Southeast Asia has the highest regional increase in frequency and intensity of human-caused fires, in part associated with drainage of the famous peat-swamp forests that are home to critically endangered orangutans. It also has the highest rates of mining and hydropower dam construction anywhere in the tropics, and huge numbers of wild animals are harvested not only for food but also international trade in traditional medicine.[25] Island biotas in this region are thus at particularly severe risk of extinction.

The grave threats facing island-dwelling species in today's world, and the specific threats to animals in human-impacted Southeast Asia, are key concerns that have shaped my career in conservation. And it was because of these two issues that one sunny summer weekend I found myself chatting with Gregory Forth and his wife in a staff coffee room at London Zoo. Forth, a British anthropologist who holds a professorship at the University of Alberta, had spent his career conducting fieldwork in the islands of eastern Indonesia. I had a specific interest with one island in particular where he had worked – Sumba, one of the largest of the Lesser Sunda Islands, in the Indonesian province of East Nusa Tenggara. I'd recently returned from fieldwork in China, and while we talked we drank some of the fancy tea I'd brought back with me. Forth suggested useful contacts in Indonesia while regaling me with wild fieldwork tales, adding fuel to my already-burning desire to get out to the Sundas. Plans began to form.

As we drank our tea, Forth talked about folklore that he'd researched across Indonesia. One of the topics he'd studied seemed particularly unusual – local stories about mythical wildmen that supposedly lived in the remote forests of many Indonesian islands, which had been the subject of one of his books, *Images of the Wildman in Southeast Asia: An Anthropological Perspective*. As soon as he left, I ordered a copy of his book online, and then thought no more about it. Little did I know what I'd just set in motion.

CHAPTER TWO
SUMBA, EAST OF JAVA

My island was now peopled, and I thought myself very rich in subjects ...

Daniel Defoe, *Robinson Crusoe*

For a biologist or anthropologist, Indonesia is almost too incredible to comprehend. It is the world's largest island nation, stretching west to east between the Asian and Australian landmasses across a distance of more than 5,000km. Its geography is unbelievably complex, consisting of over 17,000 islands (6,000 of which are inhabited), which together act as a boundary between the Pacific Ocean to the north and the Indian Ocean to the south, and which encompass 15 different seas and 25 major straits. The many islands that make up this vast island world have very different origins and histories, and the biodiversity they contain shows correspondingly fascinating and unexpected patterns of biogeography, the geographic distribution of animals and plants. And as Edward O. Wilson, one of the most important biologists of the twentieth and twenty-first centuries, has said, 'There's *nothing* more romantic than biogeography'.[1]

The islands of western Indonesia – including the huge landmasses of Borneo and Sumatra, and Java, the world's most heavily populated island – all sit on an extension of the Asian continental shelf known as the Sunda Shelf, and are surrounded by very shallow seas only 30–40m deep. The large island of Palawan in the western part of the Philippines also forms part of this continental shelf. This geographic pattern was first recognised by the English hydrographer George Windsor Earl, who referred to the region in 1845 as the 'Great Asiatic Bank'. Periodically over the past couple of million years, as the Earth's ice caps contracted and expanded during the Ice Age cycle and global sea levels rose and fell in response, the entire Sunda Shelf has been exposed above water,

creating land connections between these islands and the Asian
continental mainland for many thousands of years at a time. This
huge single landmass, which was last exposed less than 10,000
years ago, was named Sundaland by the Dutch geologist Rein van
Bemmelen in 1949 (for British readers, this is not to be confused
with Sunderland, the biogeographically less interesting city in
north-east England). Indeed, Sundaland has existed above water as
a huge south-eastern extension of continental Asia for most of the
past 800,000 years.

As a result, as noticed by Earl, the fauna and flora of the Sunda
Shelf islands are very similar to those of mainland Southeast Asia.
Different islands do show biogeographic differences. The largest
islands support many more species than the smaller islands, and small
islands are also only able to support relatively small animals, a good
illustration of the fundamental rules of island evolution.[2] Unusual
endemic species are found on some of the islands, such as the Bornean
bristlehead, a remarkable bird which looks like a crow that's been
shaved and painted in gaudy colours. This is the only member of the
bird family Pityriasidae, and only occurs in Borneo, as its name
suggests. However, biodiversity mostly shows only minor differences
between different Sunda Shelf islands, typical of the variation seen
across neighbouring parts of the same continent. Indeed, many of the
biological differences now seen between these islands and mainland
Southeast Asia are due to historical extinctions rather than local
evolution. The recent fossil record shows that tigers used to be found
across the Sunda Shelf, but they disappeared during prehistory from
Borneo and Palawan, and were hunted to extinction in the twentieth
century on Java and Bali. Whereas tigers still exist here and there on
continental Asia from India to eastern Russia, today they survive on
the Sunda Shelf only in Sumatra, where a few hundred individuals
are just hanging on in scattered small populations. Rhinoceroses
show the opposite pattern. As illustrated by the story of the floating
rhino of Bangladesh, a century or two ago Sumatran and Javan rhinos
were distributed widely across mainland Southeast Asia and the larger
Sunda Shelf islands, and bones from archaeological sites show that
they formerly occurred as far north as central China. However, today
they have vanished from continental Asia. Sumatran rhinos only

survive with certainty within a few remote rainforests in Sumatra and Borneo, and Javan rhinos are now reduced to a single remnant population in Ujung Kulon National Park on the western tip of Java. They are now two of the rarest mammals on the planet, each with global populations of fewer than a hundred individuals; they are members of what Edward O. Wilson has poignantly called the 'Hundred Heartbeat Club'.[3] Next time you hear about widely-publicised conservation efforts for rhinos in Africa, remember Southeast Asia's rhinos; they are in even greater need of awareness and attention.

However, very different kinds of animals are waiting to be encountered by the adventurous traveller elsewhere in the Indonesian archipelago. Indonesia's easternmost provinces, Papua and West Papua, cover half of the vast island of New Guinea (the eastern half of this island constitutes the independent country of Papua New Guinea). New Guinea and the neighbouring Aru Islands sit on a different continental shelf, that of the Australian continent. George Earl was again the first person to recognise this, naming the shelf the 'Great Australian Bank'. As with the Sunda Shelf, when sea levels fell periodically in the past this whole region was unified as a single exposed terrestrial landmass, usually called Sahul (or sometimes Meganesia, or the clunky composite name Australinea).[4] As a result, New Guinea's biodiversity is 'Australian' in character, with both regions containing mammal faunas dominated by marsupials, and bird faunas including cassowaries, bowerbirds and birds of paradise rather than continental Asian groups such as pheasants or woodpeckers (hence the convergent evolution of New Guinea's trioks, with their long grub-poking fingers). Until only a few thousand years ago, carnivorous thylacines – pouched marsupials that survived on Tasmania until the 1930s or even later – roamed both mainland Australia and New Guinea as the top mammalian predator. Indeed, whereas Australia is famous for its monotremes or egg-laying mammals, the duck-billed platypus and short-beaked echidna (comprising the third main living mammal group, in addition to marsupials and placentals), Indonesia is actually home to more monotreme species than any other country. As well as short-beaked echidnas, Papua and West Papua together also host three species of long-beaked echidnas, much larger animals with fewer spines, including one named after David Attenborough that's found on a single mountain range and is now as rare as a Javan rhino.

The closest distance between the Sunda and Sahul continental shelves is almost 1,500km, but the intervening region is not an empty stretch of ocean. To the east of continental-shelf Borneo is the huge ungainly island of Sulawesi, formerly known as Celebes, with its long peninsulas sprawling off in all directions; and to the east again lie the Maluku or Molucca Islands, once called the Spice Islands because of their nutmegs, cloves and mace that were so important for trade (the easternmost of the Malukus, the Aru Islands, sit on the Sahul Shelf, as this administrative grouping reflects above-water geography rather than below-water geology). Further south, the Lesser Sunda Islands (which are confusingly not on the Sunda Shelf, despite their name) act as stepping stones from continental-shelf Java and Bali almost all the way to Sahul, with an east–west chain made up of Lombok, Sumbawa, Flores, Timor and many smaller islands. George Earl described these islands as 'floating on ... unfathomable seas', and 'separated from each other by narrow channels of unfathomable depth, through which the current from the Pacific, caused by the prevalence of easterly winds, rushes with great force'.[5]

These islands are not connected to either neighbouring continental shelf, and have also never been connected to one another. Some are loose fragments of continental crust or 'microcontinents', whereas others have been built up by volcanic eruptions. Indonesia sits on the Pacific Ring of Fire, in a region where the edges of the Eurasian, Indo-Australian and Pacific tectonic plates jostle against each other and generate tremendous forces. Indonesia's volcanoes are amongst the most active in the world. The eruption of the volcanic island of Krakatau or Krakatoa in 1883 almost completely destroyed its island, was heard almost 5,000km away, and may have killed more than 120,000 people. Sixty-eight years earlier, the eruption of Mount Tambora on the island of Sumbawa was the most powerful explosion ever recorded in human history. Ash and sulphate aerosols blown into the atmosphere by Tambora's eruption caused global temperatures to drop and led to the infamous 'year without a summer' in 1816, which caused food shortages around the world and prompted Mary Shelley to stay indoors and write *Frankenstein*. And only 75,000 years ago – a blink of an eye in geological terms − the eruption of the Toba supervolcano on Sumatra was at least 12 times greater than Tambora. The Toba supereruption was the largest natural disaster known in the past two and a half million years, and must have caused catastrophic and long-term environmental impacts around the globe.[6]

It was George Earl who first noted that the faunas of the Sunda Shelf islands showed close similarities to those of mainland Asia, and the shared marsupial faunas of New Guinea and Australia reflected their shared location on a different continental shelf. However, it was another Victorian traveller, Alfred Russel Wallace, who first noticed the remarkable biogeographic pattern exhibited across the islands that lay between Sunda and Sahul. Wallace is probably most famous for independently coming up with the theory of evolution by natural selection in 1858, which came to him 'in a sudden flash of insight'[7] while suffering from an attack of fever (probably malaria) on the Indonesian island of Ternate in the Malukus, and which he outlined to a stunned Charles Darwin in a letter that Wallace later described as 'like a thunderbolt from a cloudless sky'.[8] Unlike Darwin, Wallace did not come from a privileged background, but was instead a self-taught naturalist who funded himself by collecting specimens that were sent back to London and sold by an agent. This financial necessity prompted him to develop an expert eye at spotting differences between the animals he observed during his travels.

While travelling east from Borneo and Singapore, two years into the eight-year trip that would later take him to Ternate, Wallace was forced to make a detour via the neighbouring islands of Bali and Lombok on his way to Sulawesi. He later wrote in *The Malay Archipelago*, his famous account of these travels, that 'Had I been able to obtain a passage direct to that place from Singapore, I should probably never have gone near them, and should have missed some of the most important discoveries of my whole expedition to the East'.[9] During the few days he stayed in Bali, Wallace saw birds with which he was already familiar from his time on the Sunda Shelf − weavers, magpie-robins, barbets, orioles, starlings and woodpeckers. 'On crossing over to Lombock, separated from Bali by a strait less than twenty miles wide, I naturally expected to meet with some of these birds again; but during a stay there of three months I never saw one of

them'.[10] Instead, to his surprise Wallace found a totally different bird
fauna, including cockatoos and honeyeaters. The importance of this
observation was not lost on him. 'If we look at a map of the Archipelago,
nothing seems more unlikely than that the closely-connected chain of
islands from Java to Timor should differ materially in their natural
productions,'[11] Wallace wrote, but 'we may pass in two hours from
one great division of the earth to another, differing as essentially in
their animal life as Europe does from America ... so that the naturalist
feels himself in a new world, and can hardly realize that he has passed
from the one region to the other in a few days without ever being out
of sight of land'.[12]

As Wallace knew from George Earl, Bali is the easternmost island
on the Sunda Shelf. Beyond Bali, the sea floor of the Lombok Strait
drops down to 250m, and so has remained a permanent marine barrier
throughout Ice Age sea-level fluctuations. The strong currents of the
Indonesian Throughflow, which sweep southward through the deep
strait, drastically reduce the chances that any animals or plants that are
washed out to sea will be carried across to Lombok accidentally; and
even many bird species will tend not to fly over open water, further
limiting any opportunities for colonisation. Similar biogeographic
barriers are seen on the other side of Indonesia along the edge of the
Sahul Shelf.

The geologically isolated islands between Sunda and Sahul are
therefore depauperate in comparison to the biological richness of
either continental region. Although it's possible that some of the
inhabitants of these islands (such as the tiny spider-like mite harvestmen
of Sulawesi)[13] have survived as geological passengers ever since these
continental fragments broke free from other ancient landmasses tens
of millions of years ago, the majority of their faunas have resulted
from a series of chance overwater dispersals by founders that were able
to survive risky sea crossings.[14] And these faunas are unique – they are
made up of unexpected combinations of lineages from both Asia and
Australia, which have evolved into new forms in response to the novel
environmental conditions found on these islands. For example,
Sulawesi, the largest of the isolated islands, supports a mammal fauna
comprising dwarfed buffalos, native pigs called babirusas with wrinkly
hairless skin and huge curved tusks growing out through the tops of

their snouts, and an evolutionary radiation of tiny primates called tarsiers – all of which are related to species in continental Southeast Asia – but is also home to arboreal marsupials called cuscuses. Plants also show a broadly comparable pattern of biogeographic differentiation across the islands of Southeast Asia, with the geologically isolated region between Sunda and Sahul supporting many endemic drought-tolerant species, although plant distributions are also complicated by climatic patterns and the boundary of the seasonal monsoon.[15]

To acknowledge Alfred Russel Wallace's recognition of this remarkable biogeographic region, the faunal boundary between Bali and Lombok is known as Wallace's Line, and the entire region between Sunda and Sahul is now called Wallacea in his honour. It is one of the most fantastic, endlessly fascinating and compelling places on Earth.

Very few people have heard of Sumba. Wallace himself never visited the island (the closest he got was Timor, a couple of hundred kilometres away to the east, which he visited three times between 1857 and 1861). It has a luxury resort where a distant member of the Kardashian clan got married, and where David and Victoria Beckham publicised one of their holidays in the British tabloids. London Zoo had a pair of Komodo dragons in the 1930s called Sumba and Sumbawa, because these islands lie close to Komodo in the Lesser Sundas, so the name might have become vaguely familiar at the time. The island has been romantically described as 'shaped like a horse's head with its edges painted in the whites and beiges of beaches, coral cliffs, and ocean-land ecotones'.[16]

On our way we flew from Bali in a tiny plane above the choppy Lombok Strait, and then past Tambora, rising majestically through the cloudbank to our right and dominating Sumbawa's skyline. Travelling with me was my postdoc Jen, who was hugely sensible but also hugely fun; my PhD student James; and Tim, who could actually speak Indonesian (he had a degree in it), so was clearly by far the most useful member of the team. I had travelled around Indonesia before – a few years earlier I'd spent several weeks in Indonesian New Guinea looking for critically endangered long-beaked echidnas, and I'd also explored parts of the archipelago by boat, visiting sultans' palaces and being shown lots of batik – but Sumba was new. As we stepped from the plane onto the airstrip outside Waingapu, the island's capital, a hot dry yellow landscape greeted us. It was a relief to land, as the local airline names and logos hadn't filled me with confidence: 'We strive to succeed!' 'Flying is cheap.' 'We make you fly.' And we definitely hadn't wanted to travel with Icarus Air.

We'd arrived just after the annual Pasola festival, when opponents mounted on horseback throw spears at each other, and on our way into town we drove past a huge arena full of tiny ponies trotting around. In the past the spilled blood of riders and horses would ensure a prosperous rice harvest, but today it was mainly done for tourists; riders only died accidentally these days. Despite the recent celebrations, Waingapu was pleasantly sleepy. The only other foreigners in town appeared to be a pair of dour German tourists, who frowned at us and said nothing. A zebu tethered with some string stood at the side of the dusty road next to our small hotel, its droopy tea-bag neck wobbling slowly as it chewed the cud. Although we were tired from travelling, we set out to explore and get supplies. Around the corner was a shop that seemed to sell only cakes and adult diapers, all of which were covered in dust. 'Hello mister!' yelled small boys with big grins on their faces.

Umbu was waiting for us outside the hotel that evening. He was a friend of a friend of Gregory Forth; trying to find anyone that knew anyone in Sumba who could help us had taken months. Umbu was thin and wiry, full of coiled energy, and with an intense expression. He looked very serious as we made plans for our forthcoming fieldwork, but while we talked his face suddenly broke into a mischievous grin, and his eyes twinkled. Jen and I were the two senior members of the team, he said, so we should be given respectful Indonesian honorifics for 'sir' and 'madam': Pak Sam and Ibu Jen. 'Pak Sam!' cried Umbu, and roared with laughter. With Umbu in charge, fieldwork was going to be fun.

The fossil record of Wallacea is poorly understood. Conducting scientific excavations across this vast, complex region would be an epic task, and although many of the larger islands have received some attention from palaeontologists, most of what we know comes from a small handful of rich fossil sites. Huge biogeographic gaps thus remain in our patchy knowledge about the faunas of Wallacea's geological past. The oldest mammal fossil known from the region is the partial skull of an anthracothere, a vaguely hippo-like animal, which was found in the 1960s in a deposit of Eocene age (56–34 million years old) on central Timor.[17] The specimen poses a mystery. Timor is considered to be a fragment of the ancient southern supercontinent Gondwana, but anthracotheres were otherwise only known during the Eocene from continental Eurasia, a landmass with no geological connection to Gondwana. Furthermore, Timor was submerged beneath the sea for all of the Eocene and well afterwards, until about 3.5 million years ago, and the rocks in which the anthracothere was found are apparently deep-water deposits (as far as anyone can tell from the available information about its discovery). Did a dead anthracothere wash out to sea from Southeast Asia 40 million years ago, manage to float more than 1,000km without sinking, disintegrating or being eaten, and then become fortuitously fossilised? Timor has been described as 'geological chaos'[18]; is the island even more geologically complex than we think, with this single fossil providing evidence that it actually includes a fragment of Southeast Asian continental plate wedged against a fragment of Gondwana? Or might it be that the specimen isn't from Timor at all, but actually (accidentally or otherwise) came from somewhere else? This enigmatic old fossil raises more questions than it answers – which is the best way to make science continue to move forward.

Other terrestrial vertebrate fossils from Wallacea date to within the last couple of million years or less, from the late Pliocene, Pleistocene and Holocene epochs – still an immense gulf of time, but representing the very youngest slice of the geological record, by which point all the main islands were above water and similar to how they look today. Distinctive recent fossil assemblages are known so far from Sulawesi, Flores and Timor, all exhibiting the characteristic evolutionary features of island faunas. These islands were home to endemic dwarf proboscideans and numerous species of giant rats that grew to the size of rabbits or cats – classic examples of Foster's rule.* The Wallacean proboscideans were all stegodons, elephant-like beasts that diverged from the ancestors of African and Asian

---

* Another endemic species of now-extinct giant rat is now also known from the Wallacean island of Alor.[19]

elephants around twenty million years ago and grew to comparable sizes in mainland Asia, but in some cases shrank to less than half a tonne[20] – only a bit bigger than a tapir or wild boar – once they became isolated on islands. Although the initial discovery of stegodon fossils on Flores in the 1950s was described as 'a heavy blow for the Wallace Line',[21] this is just another example of the excellent overwater dispersal ability of proboscideans, which are more at home in water than most other mammals.

The consistent pattern of evolutionary dwarfing on islands, along with evidence for repeated overwater colonisations by different proboscideans, is demonstrated on Flores across a series of different fossil assemblages that span the last 900,000 years. The oldest fossil beds contain a stegodon that had already become an island dwarf, but this animal seems to have died out 100,000 years or so later (possibly due to catastrophic volcanic activity), and is replaced by a markedly larger stegodon, which probably represents a new overwater arrival that hadn't shrunk yet. The youngest fossil beds, which are only about 50,000 years old, contain a third, dwarf stegodon that almost certainly represents a descendant of the second arrival that had shrunk once again over time.[22] The islands also supported giant reptiles: giant tortoises on all three islands, Komodo dragons on Flores and a related huge predatory lizard on Timor, and another weird large monitor lizard on Flores called *Varanus hooijeri* with very distinctive blunt rounded teeth, possibly for crunching either snails or hard seeds and fruit stones. These animals are well suited to surviving in resource-poor island ecosystems, because their metabolisms don't require as many resources as those of warm-blooded mammals. As well as Komodo dragons, prehistoric Flores also had another terrifying predator – a giant marabou stork, which grew to almost 2m tall and was probably flightless, and had a huge lethal stabbing bill for feeding on giant rodents and anything else that came its way.[23]

These insights into the recent palaeontological past provide a unique new perspective for understanding Wallacea's biodiversity and how it has changed through time. However, the region exemplifies Darwin's description of the fossil record as 'a history of the world imperfectly kept … of this history we possess the last volume alone …

Of this volume, only here and there a short chapter has been preserved; and of each page, only here and there a few lines.'[24] For most of the Wallacean archipelago, and even for many of the largest islands such as Sumbawa and Lombok, we still have no fossils at all.

Sumba was slightly different, and even more tantalising. In August 1978, a survey team comprising archaeologists and geologists from Java visited the island. As they returned to Waingapu from a visit to an ancient urn-burial site on the east coast, the group stopped briefly at a small dry valley, the remains of an ancient riverbed, which ran beside the road close to the kampung or small village of Watumbaka. During this brief stop, one of the geologists made a hugely important chance discovery – the fossilised mandible or lower jaw of a stegodon was sticking out of the hard calcified limestone of the long-dead riverbed. Night was falling and the survey team rapidly started to excavate the fossil, although 'the only illumination available was provided by headlights of a jeep.'[25] The mandible was taken back to the Bandung Institute of Technology on Java, where it was described as a new species endemic to Sumba, *Stegodon sumbaensis*. But after this exciting episode of nocturnal fossil-hunting adventure – which took place when I was one year old – nothing else had been reported. It was hard to tell whether any other palaeontologists had even visited Sumba again.

Why would a conservation biologist be interested in fossils? It's a good question, since the goal of conservation is to prevent today's threatened species from becoming extinct, whereas the fossil record instead contains the remains of species that are already long gone. However, the past – and especially the recent past – can provide hugely important insights into today's biodiversity and how to protect and maintain it, which cannot be obtained from anywhere else. Because so many island species are now extinct, the recent fossil record often provides the only source of information on past diversity and environmental states before humans arrived and disrupted everything. These unique snapshots allow us to understand how ecosystems functioned in the past, how much they have changed from historical baselines, and which species have been more vulnerable or resilient to different kinds of change through time. In some cases, modern species are also represented in the fossil record but in locations or ecosystems where we would otherwise have no idea they could survive, providing evidence for past population declines and revealing they can tolerate broader ecological conditions than expected.

Flores used to support a large(ish) herbivore guild of stegodons and giant tortoises that have vanished only recently in geological and evolutionary terms. How many of the island's modern plants co-evolved with these extinct browsers and grazers; and now that their seed and fruit dispersers are gone, are they less able to survive without specific conservation attention? Were these lost herbivores actually so-called keystone species or 'ecosystem engineers' that modified the very structure of their environment through stripping and trampling vegetation and distributing nutrients across the landscape in their dung, making the soil more productive than it is today? Komodo dragons still survive on Flores, but the fossil record reveals that a close relative also once occurred on Timor – so should we consider introducing this now-threatened species to other islands that have lost their top predators, as a step towards restoring the region's damaged ecosystems? And ultimately, in the face of the constant environmental and biological change seen over time (for example, the different types and sizes of stegodons in successive fossil deposits on Flores), how do we even define what constitutes a 'natural' ecosystem, or an objective restoration target to aim for? These are all hugely important questions with direct relevance for conservation today, and with no easy answers. However, we wouldn't even realise we needed to think about them without an understanding of the past.

In some cases the link between the fossil record and modern-day biodiversity is even more direct, with species first described from fossilised remains later surprising everyone by turning up alive and well. A dramatic example is described by Tim Flannery in his fantastic account of biological adventuring in New Guinea, *Throwim Way Leg*, a book that inspired me hugely as a student. A very unusual fruit bat was described in 1975 from 12,000-year-old fossil bones; it was named Bulmer's fruit bat in honour of Susan Bulmer, the archaeologist who discovered its remains in a cave in the Papua New Guinean mountains. It was a remarkable animal – the world's largest known cave bat, it uniquely lacked any front teeth. Two years later, an anthropologist documented a bat hunt 400km away to the west, in a vast vertical cave called Luplupwintem that drops directly down for more than 300m. Shotguns and nylon rope had reached this remote region for the first time, and a brave villager climbed down into the cave and shot thousands of bats for a huge feast. A skull from the slaughter made it back to the University of Papua New Guinea (the rest of the specimen had been eaten by a village dog that snatched it off a dinner table) and was identified as Bulmer's fruit bat, but by this time the cave's bat colony had apparently been hunted to oblivion. Had the species gone extinct for a second time? Luckily not; when Flannery visited Luplupwintem in 1992, he was able to catch a bat by perilously setting a mist-net between trees hanging above the huge mouth of the cave. 'We looked in amazement at the indignant face of this bat. Its incisors were missing. In our grasp was an animal once thought to have become extinct at the end of the last ice age, some 12,000 years ago. We hugged each other with joy—after eight years of field work together in rugged western Papua New Guinea we had rediscovered Bulmer's Fruit-bat!'[26]

The remarkable story of Bulmer's fruit bat is far from unique. The South American bush dog and Chacoan peccary were both described from fossils, and thought to be extinct before surprising scientists by popping up in the flesh; the bush dog's scientific name, *Speothos*, means 'cave wolf' in reference to its fossil origins. Such remarkable discoveries haven't been restricted to remote tropical regions, either. In 1846, Richard Owen (the famous Victorian anatomist who coined the word 'dinosaur') reported a 126,000-year-old fossil skull of a small whale found three years earlier in the Lincolnshire fens of eastern England, describing it as a new extinct species that he named the 'thick-toothed grampus'. Fifteen years later, freshly dead individuals of the same animal washed up in Denmark, and the species, today called the false killer whale, is now known to be widespread and relatively common across the world's oceans. These animals are all living fossils in the most literal sense.*

---

* There are several other definitions of the term 'living fossil', a concept first coined by Darwin, including a species or lineage that has remained (relatively) unchanged for millions of years, or an evolutionary relict of a once-large radiation. The term is often avoided by evolutionary biologists due to this diversity of definitions.

Although living fossils such as thick-toothed grampuses have occasionally turned up only a few miles from the doorstep of western scientists, the potential for exciting new fossil discoveries with direct significance to current-day biodiversity is much greater today in the world's less-explored far-flung corners. Sumba is a perfect example of such a place. The only decent paper on Sumba's living mammals was written in 1928 by the Dutch zoologist Karel Dammerman, who considered the island's fauna 'to be extremely poor'.[27] Other than a few bats, all the land mammals recorded by Dammerman were widely distributed across Indonesia and had probably been introduced to Sumba in the past by humans. But was this really all there was to know about Sumba's mammals?

Sumba is only 50km south of Flores, and on a clear day north of Waingapu you can see Flores on the horizon. The strong currents of the Indonesian Throughflow, one of the major crossroads of the world's ocean circulation system, pull huge amounts of water southward past Flores and other islands towards Sumba, potentially transporting any animals that get swept into the sea and making them overwater colonists. Sumba has never been connected to any other Indonesian islands; it is an ancient continental fragment, although whether it originally broke free from Australia or Asia tens of millions of years ago is unclear. Its rocks consist almost entirely of alternating shallow-water and deep-water marine limestones (with a few volcanic rocks poking out here and there), revealing that the island has been underwater for most of its recent geological history – and with no out-of-place anthracothere fossils to complicate the picture, unlike on Timor. Any animals washed up by accident after Sumba rose above the sea around one million years ago would thus have been isolated in a novel environment, and could potentially have evolved into something new, like the unique stegodon found at Watumbaka all those decades earlier.[28] For birds, the region's best-studied animal group, Sumba has a high level of endemism, with 14 living endemic species recognised at the time of our visit (although how distinct they all are from populations on neighbouring islands, and whether other members of Sumba's avifauna might also be endemic, continues to be debated in the best scientific fashion).

And the abundance of limestone across Sumba provided further cause for excitement. Limestone is often full of caves that have been dissolved out of the soft rock by slightly acidic rainwater and groundwater. These caves usually have a short lifespan in geological terms, being prone to eventual erosion or collapse. However, they are sheltered from the weathering effects of rain, sun, and fluctuating temperatures, and can accumulate ancient bones and other remains across time periods of tens or hundreds of thousands of years. As well as old riverbeds, caves are among the best places to find Pleistocene and Holocene fossils. They have been vital sources of information about prehistoric faunas ever since William Buckland excavated Kirkdale Cavern in Yorkshire in 1821 and discovered the remains of an Ice Age hyena den. Archaeological excavations in caves and rockshelters have also yielded remains and cultural artefacts of prehistoric humans ever since cave archaeology was pioneered by Édouard Lartet, when he discovered Cro-Magnons in southern France in the 1850s.

So one of the reasons we had come to Sumba was to look for fossils: to try to gain further insights into what animal life had been like in the past on this fascinating island, and to see whether fossil discoveries could tell us something new about environmental conditions today – with the possibility of maybe even finding some living fossils of our own. Our work was an official collaboration with Indonesia's national geology and palaeontology museum in Bandung, and Umbu was the ideal professional to work with as our project partner on Sumba; he was an expert caver employed by a local non-governmental organisation that mapped underground water sources, to provide support for rural communities in need of basic resources. If anyone knew where to find caves that might contain bones on this island, it was Umbu. Not everyone back home had quite grasped what we were aiming to do, however. At the travel clinic getting his final jabs, Tim had explained to the nurse that we were planning to look for fossils of long-dead animals in caves. The nurse looked cautious and disapproving. 'You be very careful,' she had said. 'Sometimes those animals are only *pretending* to be dead.'

Other than Sartono's paper describing the 1978 stegodon discovery, though, we weren't sure where to look. Watumbaka was

the obvious place to start in the absence of information about other
fossil sites. We set off east out of town, wearing the team t-shirts
that Ibu Jen had made for us, emblazoned with the immortal
geological logo, 'The Holocene – We're All In It Together'. Soon
Waingapu was behind us and we drove along a road that ran parallel
with the sea a mile or so to our left. A brahminy kite – a beautiful
bronze bird of prey with a striking white head and belly, found
widely across Southern Asia and Australia – circled over the tiny
huts dotted along the side of the road, looking for chickens. The
landscape was barren; everywhere was just rocks, parched grass,
scrub. Cactuses, introduced at some point from the New World,
grew through abandoned car tyres and along the edges of the thin
stick-and-wire fences separating the huts by the roadside. This
landscape was not what might be expected in tropical Southeast
Asia, but the slow geological journey of the Sumba microcontinent
had brought it to a very dry part of the world. The moist monsoon
winds that carry rain to the tropical forests of Borneo and Sumatra
have dumped almost all their water and exhausted themselves by
the time they reach the Lesser Sundas, and dry trade winds sweep
up from Australia to further regulate the local climate. Sumba and
its neighbouring islands are in a natural rain shadow, and their
faunas and floras have become adapted to a very dry lifestyle. The
north coast of East Sumba through which we were driving is the
driest region in all of Indonesia, and is more similar to the arid
landscapes seen across great swathes of Australia.

We stopped near Watumbaka and explored an erosion gully carved
into the grey soil, full of huge fossil clams that had lived here millions
of years ago when Sumba was just a coral reef under the sea. We found
the dry riverbed by the road just outside the tiny hamlet and wandered
around, prospecting for more recent fossils. The sun was low in the
sky, making the rocky rubble we were exploring glow orange and
pink; even the light felt different in this harsh, odd landscape. There
was no sign of another stegodon.

We drove back west to Waingapu as the sun sank beneath the sea
on the horizon. Back at the hotel, Umbu sat with us as we discussed
what to do next. I remembered my copy of Gregory Forth's book,
from which I'd been reading extracts to the rest of the team over the

past few days as we'd travelled through Indonesia; it had contained a section about Sumba. As we finished making fieldwork plans, I thought of something else to ask. 'Umbu, have you heard of the mili mongga?' I saw sensible Ibu Jen smile and roll her eyes. But Umbu didn't laugh. 'Yes, Pak Sam!' he frowned in thought. 'I have heard people talk about it.' I had to ask more. 'Do you know anyone who might be able to tell us about it?' Umbu promised that he'd ask around and see what he could find out. I was definitely not prepared for where that question would take us.

CHAPTER THREE

GLUTTON-GRANNY

About forty years ago some anthropologists believed that the first human being was a pygmy and that the pygmies of today are the relics of that first human type. Sometime later, a paleontologist discovered a Tertiary pebble on the beach of Antwerp which showed on the surface a strange system of parallel curved lines. He interpreted this pattern as a print left by the tip of the big toe of a man who once had taken a sun bath on the beach. The man's stature, he claimed, could not have exceeded 1 yard, calculated from the size of the area of the impressions of the big toe. Nobody took this "evidence" of the dwarf theory seriously.

Franz Weidenreich, *Apes, Giants, and Man*

The geophysical barriers that isolate islands, and that have promoted the evolution of spectacular insular biodiversity, have also had a profound effect on the local development of island-dwelling peoples. Many things have been written about the supposed characteristics of island nations, often reduced to the level of stereotype. For instance, the perceived British 'island mentality' – of either proud independence or geographic obliviousness – is regularly illustrated with reference to a famous (but probably apocryphal) newspaper headline: 'Fog in Channel – Continent Cut Off'. However, human populations on islands the world over have diversified into a complex variety of unique and fascinating local cultures and cultural identities as a result of their isolation, as well as through other factors such as the differing origins of their colonists. Indeed, in much the same way that islands have been crucial to evolutionary biologists for understanding how biodiversity is generated, islands have often been treated as discrete 'manageable microcosms' by anthropologists trying to appreciate the scope of human behaviour and to determine the patterns, processes and histories of many social and cultural characteristics, values and worldviews.

The incredible range of cultural diversity shown by human communities on islands is seen clearly across the vast, bewilderingly complex archipelago of Indonesia. The country's statistics are staggering. All the larger islands support distinct local cultures and traditions, and different regions vary both in their primary state-recognised religion and also in other Indigenous belief systems; Indonesia contains ancient Buddhist shrines on Java, Hindu water temples on Bali, and the world's largest Muslim population. There are more than 700 local languages throughout the country, around a tenth of the languages in the world, spoken by 633 recognised ethnic groups. However, most people are also able to speak 'standard' Indonesian or Bahasa Indonesia, a lingua franca based largely on the language spoken around the Straits of Malacca, allowing communication and trade between all these different peoples. The country's national motto is 'Unity in Diversity'.

Although Sumba has an area of only around 11,000km², about half the size of Wales, it has seven native languages and a very distinctive local culture. Although they are 'officially' registered as Christian or Muslim, much of the population is animist, with a complex belief system called Marapu based upon cosmological interactions with deceased ancestors and nature spirits. Sumba is one of the last places on Earth where people still construct megalithic tombs, using huge limestone slabs to inter their dead in structures resembling the ancient dolmens of Neolithic Europe. The traditional dwellings of living Sumbanese people are equally striking, with hugely tall central towers that jut above the rest of the thatch; this peak is the sacred part of the house, where the ancestors are considered to reside.[1]

The huge diversity of cultural evolution across island communities isn't the only form of evolution seen in people on islands. Everyone alive today belongs to the same species, sharing a common ancestor who lived only a few tens of thousands of years ago. However, human populations that became historically isolated on islands – especially in the millennia before overwater travel was a regular occurrence – have also been exposed to different environmental pressures, which have sometimes led to an accumulation of biological as well as cultural changes.

One of the first changes that can affect island communities is lack of immunity to rapidly-evolving outside diseases to which these sheltered communities have not been exposed. The effects of novel pathogens can be exacerbated by the limited genetic variation often present in many small island communities, which is associated with lower variation in immune response.[2] As a result of this epidemiological isolation, when people on remote islands have come into contact with visitors from larger, more disease-riddled human societies the results have often been catastrophic. Such outcomes have even occurred during recent history within the British Isles. The tiny Scottish archipelago of St Kilda, situated 65km into the Atlantic past the main islands of the Outer Hebrides, was inhabited for millennia until the entire remaining population was evacuated in 1930. The islands were remote enough to support unique subspecies of wren, field mouse and house mouse, as well as a tiny endemic dandelion, and this isolation also worked its effect on the local human population. In the words of Charles Maclean in his famous account of the archipelago's history, the people of St Kilda developed 'certain peculiar physical characteristics due to the influence of the environment'.[3] When outsiders periodically visited St Kilda, the locals would be struck down with an illness they called *cnatan-na-gall* or 'the stranger's cough', which would spread through the entire community and manifest as an aching of the jaw and limbs accompanied by headaches, depression, a cough and other flu-like symptoms.[4] Outsiders, who called the disease 'the boat-cold', often dismissed it as psychosomatic or caused by the harsh environment of the islands themselves. After his visit to St Kilda, Samuel Johnson considered that 'the evidence was not adequate to the improbability of the thing', joking that 'the steward always comes to demand something from them; and so they fall a coughing.'[5] However, six locals died of the stranger's cough between 1830 and 1846,[6] and when eight St Kildan families emigrated to Australia in 1852, more than half were killed on the voyage by measles they probably picked up in Glasgow or Liverpool.[7]

This sort of tragic outcome has unfortunately been the rule rather than the exception throughout history when island communities received visitors. Within fifty years of Columbus's arrival in the Caribbean in 1492, the native Taino people had become virtually extinct, largely due to the introduction of smallpox, measles, typhus, and bubonic and pneumonic plague from Renaissance Europe.[8] Lethal epidemics of infectious diseases, typically a first wave of dysentery followed by smallpox, measles and influenza, swept through the archipelagos of the tropical Pacific during the sixteenth to nineteenth centuries following first contact with European explorers, causing massive depopulation across this vast region.[9] On Hawai'i, contemporary observers reported that mortality was so high that 'the living were not able to bury the dead'[10], and it's suggested that numerous reports of freshwater shark attacks in Fiji during the early nineteenth century were the result of human bodies being discarded in the island's rivers, which attracted bull sharks upstream from the sea.[11] Robert Louis Stevenson, who spent his final years in the South Seas, described 'death coming in like a tide'.[12] The Indigenous population of Sumba was also ravaged by epidemics of smallpox during the nineteenth century.[13]

But could humans also undergo more extreme evolution on islands, resulting in physical change as well as immune resistance? People living on isolated islands have sometimes ended up with unusual genetic characteristics 'fixed' in their populations following historical catastrophes, after which the entire community was descended from a handful of survivors who by chance had a particular heritable condition. An example of this phenomenon, the high prevalence of the unusual genetic disorder achromatopsia on the Micronesian island of Pingelap, is described by neurologist Oliver Sacks in *The Island of the Colour-Blind*.[14] And although primates don't occur naturally on many islands, there are enough examples of island primates, from Madagascar's lemurs to Sri Lanka's slow lorises and the endemic tarsiers of Sulawesi, to demonstrate that these mammals also follow the characteristic pattern of the island rule – small-bodied species tend to get larger, and large-bodied species tend to get smaller.[15] We humans, although unique in many ways, are also primates, at the very large end of the group's size range. So are tiny island humans, like the Lilliputians in *Gulliver's Travels*, found only in fiction? Could human populations that become isolated on islands also undergo a reduction in body size in accordance with this general evolutionary trend?

The answer is ... possibly. Various pygmy peoples – ethnic groups who are markedly shorter in height than the global human average – exist around the world, although the term 'pygmy' is nowadays considered potentially derogatory. Several of these ethnic groups live in continental settings, such as the Efé and Mbuti of the African Congo. Numerous theories exist to account for the evolution of diminutive body size in these human groups, none of which are supported by very firm evidence or are likely to be applicable to all small-bodied peoples. Does small size confer resilience to starvation, or an improved ability to move through dense rainforest undergrowth or maintain a constant body temperature? Is it an evolutionary effect of reduced calcium uptake caused by vitamin D limitation from living under a forest canopy? Or does it reflect an evolutionary shift in the onset of puberty, to allow earlier reproduction in communities that experience increased natural mortality? Some large mammals also decrease in size in rainforest habitats – contrast African savanna elephants and giraffes with their much smaller rainforest-dwelling relatives, forest elephants and okapis. At the moment our understanding is still rather speculative; these various theories might not be mutually exclusive, and different factors might be responsible for local evolution of small body size in different parts of the world. It's clearly complex.[16]

Evidence for island dwarfing in humans has been further clouded by erroneous claims. In 2008, Lee Berger and colleagues published a paper based on archaeological excavations they conducted in two caves on rocky islets in Palau, a tiny island nation in the western Pacific. They reported the discovery of numerous fragmentary human remains, between 1,500 and 3,000 years old, which they suggested were evidence of a previously unknown small-bodied human population, maybe 30–45kg in adult body mass. Although these fragmentary remains were definitely attributable to our species, *Homo sapiens*, they also appeared to show extremely primitive morphological features, such as enlarged brow ridges, tiny brains, and no chins. Was this evidence of an isolated island population of dwarfed prehistoric humans?[17] No, said other archaeologists who worked in the region and revisited Berger's caves. Other sites on Palau contained the remains of people with typical skeletal proportions, and there was strong evidence for cultural continuity across the archipelago throughout prehistory – making it highly unlikely that both tiny humans and people resembling modern-day Palauans could have coexisted side by side for millennia. Extrapolating body mass and height from fragmentary specimens was criticised, especially because bones from other sites were of normal height but were quite slender, suggesting the potential for misinterpretation. It was also possible that some bones were from children rather than adults. Even more damningly, the apparent brow ridges turned out on closer examination to be precipitated calcium carbonate that could be flaked off.[18]

Some modern human populations on islands are very small-bodied, raising the possibility that they could have been shaped by the evolutionary process of island dwarfing. The inhabitants of the Andaman Islands, an isolated archipelago in the Bay of Bengal, were mentioned by Marco Polo, who probably passed within sight of the islands in 1292 and reported that 'all the men of this island have heads like dogs, and teeth and eyes like dogs; for I assure you that the whole aspect of their faces is that of big mastiffs.'[19] Polo's description was make-believe, and magical dog-headed people or cynocephali were thought to inhabit many distant lands by classical and medieval writers. However, the Andamanese are strikingly different to the people of mainland India, Bangladesh or Myanmar. They are small, slender people, well below 1.5m in height as adults, and with very dark skin and tightly curled hair. They historically practised a hunter-gatherer lifestyle, and were among the very few people who lacked the knowledge of how to make fire when first encountered by Europeans.[20] In common with the sad fate of many other island peoples, the Andamanese were almost obliterated during the eighteenth and nineteenth centuries by measles, influenza and other introduced diseases (as well as by wider social disruption, including factors such as alcohol). Their superficial similarity to small-bodied African ethnic groups such as the Efé and Mbuti led nineteenth-century anthropologists to assume that these geographically distant peoples might share a recent ancestry, with the Andamanese possibly descended from slaves 'escaped either from some Arab slave-ship carried out of its course by adverse winds, or from a slave-ship wrecked on the Andamans on its way to the Portuguese settlement in Pegu'.[21] However, genetic studies have shown that these isolated islanders are instead closer to Indigenous groups from New Guinea and Melanesia, and probably represent the descendants of the first wave of modern humans to spread into Southeast Asia between 50,000 and 70,000 years ago.[22]

The Andamanese are one of several ethnic groups scattered across South and Southeast Asia with small stature and very dark skin. Most are also hunter-gatherers and many live on islands, including several islands in the Philippines. These Indigenous groups are thought to be the last representatives in tropical Asia of the first expansion of modern humans out of Africa.[23],* They have historically been called 'Negritos', but are now sometimes referred to by anthropologists as the 'Qata' or the 'First Sundaland People'.[24] Most other Southeast Asian peoples are instead descended from a much more recent population expansion of Austronesian language–speaking Neolithic farmers who spread out from Taiwan between 5,000 and 7,000 years ago, a demographic event described by Jared Diamond as the 'Chinese steamroller'.[25] There are other small island-dwelling people in Southeast Asia too. Flores, the island just north of Sumba, is populated by Austronesian people, but these include a community known as the 'Rampasasa pygmies', most of whom are below 1.5m in height.[26]

So, there are definitely small-bodied people on islands. But is this evidence for the evolutionary process of island dwarfing also occurring in humans? Not necessarily. In marked contrast to regular body-size trends seen across other island mammals, most Indigenous peoples on islands are not reduced in size compared to continental peoples. Conversely, many ethnic groups with short statures don't live on islands; in addition to the Efé and Mbuti in the Congo, several of the surviving Qata groups, including the Maniq of southern Thailand and the Orang Asli or 'Original People' of Peninsular Malaysia, actually live in mainland Southeast Asia. It's possible that the relatively recent prehistoric arrival of many island peoples means that sufficient time has not yet elapsed for a decrease in body size to occur – the same reason why the newly arrived stegodon in the intermediate fossil beds

---

* However, it remains debated whether all of these groups have a similar long shared evolutionary history, or whether their similar phenotype is instead the result of evolutionary convergence, or a mixture of both.

on Flores was still large-bodied – and that only the ancient Andamanese and other insular Qata ethnic groups have inhabited their islands long enough to allow substantial morphological evolution. However, reduced body size in human populations might instead be associated with other specific factors, such as tropical rainforest environments, which incidentally happen to be present on some islands. Or, as evidenced by the geographic distribution of Qata ethnic groups across Southeast Asia, might small-bodied peoples on islands have originally changed in size on the mainland, and then colonised islands when they were already small? Are different island populations small for different reasons, or for combinations of some or all of these reasons? The jury is still out.

And just because non-human primates follow the island rule when they become isolated on islands, why should we expect humans to do the same? Unlike other terrestrial animals, humans are not solely reliant upon terrestrial food sources. Long before the advent of agriculture provided a further means to regulate and maintain dietary intake, people living on islands the world over have also exploited marine resources as a major source of food from both nearshore and offshore environments – fish, molluscs and crustaceans, and often also marine mammals, birds and reptiles, jellyfish, sea cucumbers, sea urchins and seaweed. The ecological restrictions that preferentially select for smaller body sizes in large island animals have therefore not necessarily applied to resourceful humans.

Indeed, one of the world's most spectacular and unusual examples of marine resource use takes place on Sumba and other islands in the Lesser Sundas – the annual harvest of nyale worms. These brown, blue and green marine worms, also known as palolo worms elsewhere in the tropical Pacific, have a remarkable reproductive cycle. In order to mate they grow a detachable mobile tail full of eggs and sperm called an epitoke, which also contains an eyespot that can detect light. When the moon wanes in February, the worms release their epitokes, which propel themselves towards the surface in huge mass spawning events, break up into small fragments and 'dissolve into a yellow creamy matter' in the words of a nineteenth-century observer, freeing eggs and sperm into the water.[27] Although they sound somewhat unappetising to a western palate, the nyale worm epitokes constitute an important delicacy and their predictable appearance marks an important part of the calendar for the people of Sumba and other islands, associated with festivities and feasting and setting the date for the annual Pasola competitions. Many myths about the creation of Sumbanese society are related to the sea worms, and specifically contrast the lack of food available on land with the plentiful bounty provided by the sea. 'If you don't know where to start,' an old Sumbanese man advised the anthropologist Janet Hoskins, 'start with the sea worms. That is where we start ourselves.'[28]

However, *Homo sapiens* wasn't the first species of human to live in Southeast Asia. Other hominin species – upright primates more closely related to us than to the other living great apes, although often not our direct ancestors – spread out of Africa before modern humans evolved, and were present in many Asian ecosystems for over a million years. The first ancient Asian human fossils were discovered in 1891 on Java by Eugène Dubois, a Dutch surgeon inspired to search for the 'missing link' by the famous nineteenth-century German zoologist Ernst Haeckel (who speculated that humans had originated via intermediate ape-men or 'Pithecanthropi' in tropical Asia, contrary to Darwin's suggestion that humans evolved in Africa). Dubois named his monumental discovery *Pithecanthropus erectus* in honour of Haeckel's insightful suggestion, although 'Java Man' is now reclassified as a member of our own genus (*Homo erectus*) and regarded as a regional side-branch rather than part of the direct line of descent of *Homo sapiens*. Dubois' Java Man finds, a skullcap, tooth and femur, are on display in Naturalis, Leiden's magnificent natural history museum. This is a far cry from their earlier history following their arrival in Europe; increasingly suspicious of differing interpretations by other palaeontologists about the fossils and their place in human evolution, Dubois hid them under his bed for years.

   Homo erectus was a part of the Sunda Shelf mammal fauna for a long time. The original Java Man specimens, found on the bank of the Solo River at Trinil, are estimated to be between a million and half a million years old. Other *Homo erectus* remains subsequently discovered on Java are up to 1.3 million years old.[29] The youngest-known specimens, a series of skullcaps excavated in the 1930s at another site on the Solo River called Ngandong, were vividly depicted by the Czech palaeoartist Zdeněk Burian, impaled on forked branches driven into the sand as 'relics of Javan Neanderthal head-hunters'.[30] Burian's painting was rich with menace and fascination when I first encountered it as a child, in an old book on human evolution I was given one Christmas. This striking image was inspired by the colourful interpretation of the site by palaeontologist Ralph von Koenigswald, who thought the Ngandong skulls provided evidence that Java Man had eaten the brains of his foes to acquire their wisdom and skill. This speculative interpretation is far more dramatic than the actual depositional process by which the skulls ended up being buried at Ngandong – probably through natural water-borne accumulation of different parts of carcasses in different parts of the river. Previous studies have suggested the skulls might be as young as 30,000 years old, supporting an overlap in time and potential interaction with *Homo sapiens* in Java, but recent reanalysis indicates they are actually between 117,000 and 108,000 years old.[31]

Although separated from us by well over a million years of evolutionary time, *Homo erectus* apparently shared some surprisingly modern behaviours with *Homo sapiens*. In 2015, a paper in the top scientific journal *Nature* revealed that fossilised freshwater mussel shells from the Dubois collection in Leiden, found alongside the original Java Man bones at Trinil, were engraved with geometric patterns. The shells were at least 300,000 years older than other known human engravings, and suggested that their carvers had cognitive abilities closer to our own than previously expected.[32] However, these ancient humans were still far from possessing the diverse range of sophisticated, flexible and adaptive behaviours that define our species. Around 50,000 years ago (although maybe somewhat earlier at some sites in Africa), evidence of a characteristic suite of new behaviours suddenly appears in the archaeological record – a 'creative explosion' of innovation, which might have been the event that propelled modern humans out of Africa and around the world. These behaviours include a huge increase in modes of artistic expression, ornamentation and ritual, together with different means of organising living spaces and new ways of working bone, antler and stone.

The last of these innovations was associated with a wide new range of methods for catching food. In particular, while there is now evidence that Neanderthals sometimes exploited marine resources close to shore (notably the presence of abnormal bony growths known as 'swimmer's ear' in some Neanderthal ear canals, associated with regular immersion in cold water), the 'creative explosion' is the first time that fishing apparently became a regular human activity. The world's oldest known fish hook was actually discovered on Timor, the next island along from Sumba. The hook itself dates from between 23,000 and 16,000 years ago, but was found alongside an archaeological deposit up to 42,000 years old which contained bones of offshore pelagic fish such as tuna. These fish can only be caught using advanced fishing methods, suggesting that similar hooks have been around for at least that length of time.[33]

Fishing, whilst constituting an ancient human behaviour that goes back tens of thousands of years, thus substantially postdates the existence of *Homo erectus*. Modern humans might be freed from the general evolutionary trends experienced by animals on islands, because they can exploit rich marine resources instead of being limited to terrestrial resources; but *Homo erectus* and other ancient hominins wouldn't have been able to do this. If any populations of these early humans became isolated on islands in Southeast Asia, might they therefore also have evolved smaller bodies in accordance with the island rule?

There have been several intriguing suggestions that some hominins in island Southeast Asia were weird in size. As early as the 1940s, some of the incomplete jaws and skulls found alongside *Homo erectus* remains in Java were interpreted as belonging to a larger-bodied species of ancient human. This hypothetical hominin was even given the formal scientific name *Meganthropus palaeojavanicus* ('The Great Man of Ancient Java') by Ralph von Koenigswald, a few years after his work on the Ngandong crania, on the basis of a particularly robust mandible fragment 'exceeding in size and massiveness all that is known or could ever be expected of a hominid'.[34] Von Koenigswald later wrote that he kept this fossil in his pocket for at least a week, constantly checking and re-checking the specimen because he couldn't quite believe it.[35] His colleague, Franz Weidenreich, called it 'the most unexpected and exciting'[36] fossil human find ever made. Weidenreich was so enthused about the specimen's importance that he actually rushed out a publication describing it during the Second World War, assuming that his colleague in Java might have died under Japanese occupation but 'sure of von Koenigswald's consent'.[37]

The jaw was as large as a gorilla's. If it had human-like proportions, the mysterious *Meganthropus* has been speculatively reconstructed as maybe reaching eight feet tall. Weidenreich considered the specimen as supreme evidence of a hitherto-unexpected evolutionary pathway: 'I believe … that the human line leads to giants, the farther back it is traced. In other words, the giants may be directly ancestral to man.'[38] However, the reality of this supposed island giant has been challenged by most subsequent palaeontologists, who preferred to interpret von Koenigswald's jawbone fragment and other potential *Meganthropus* bits and pieces as just heavy-set chunks of normal *Homo erectus*. More recently, detailed analysis of the morphology and internal structure of *Meganthropus* jaw fragments instead suggests that this mystery animal might actually have been a kind of non-human giant ape. *Homo erectus* thus seems to have lived in an unexpectedly diverse prehistoric ecosystem, which contained not only orangutans but also other large primates.[39]

A large-bodied primate such as a hominin wouldn't be expected to become even larger in size under isolated island conditions, so reinterpretation of *Meganthropus* as a large ape rather than a giant mystery man-beast solves an evolutionary inconsistency. Although Java is currently an island, it's also the eastern tip of the continental shelf of Sundaland and so formed part of mainland Southeast Asia for much of the period when *Homo erectus* existed. Conversely, the geologically isolated and biogeographically weird islands of Wallacea further east, once home to giant rats and tiny elephants, are perfect natural laboratories where evolution might also reshape any hominins that somehow made their way there. However, these remote islands were only colonised through chance events, and by a small subset of the terrestrial faunas of mainland Asia and Australia that managed to survive extremely rare and risky sea crossings. The big question, then, is: before our species spread out across the islands of Southeast Asia a few tens of millennia ago, did any other hominins also make it to Wallacea?

For several decades there have been indications that some sort of hominin pre-dating *Homo sapiens* had existed in Wallacea, and might have been there for many hundreds of thousands of years. Indeed, the first discovery of stegodons on Flores was accompanied by evidence of ancient stone tools. In 1957, after being tipped off by the local Raja about a large bone found on a hunting trip, a Dutch missionary called Theodorus Lambertus Verhoeven discovered fossilised stegodon remains near the abandoned village of Ola Bula. In the same deposit, Verhoeven found primitive stone flakes and blades typical of the archaic Palaeolithic (Old Stone Age) tools made by hominins such as *Homo erectus*.[40] Similar tools were subsequently found at several other sites on Flores, Sulawesi, Timor and Roti, in sediments dating to more than 800,000 years old.[41]

Verhoeven and other researchers interpreted these ancient tools as evidence that *Homo erectus*, already known to have lived on Java during the Pleistocene, had somehow made the overwater journey eastward across Wallace's Line. Verhoeven's claims of ancient human presence in Wallacea were initially ignored, but as evidence accumulated, it prompted many anthropologists to suggest that this journey could only have taken place through deliberate construction of boats and intentional – daring! – seafaring.[42] The cognitive and cultural abilities of ancient Asian hominins – including whether they could communicate complex ideas using language – must have been greatly underestimated, these anthropologists argued, if they had managed to develop maritime technology so long ago. Had the apparent 'creative explosion' 50,000 years ago actually been a much longer-term, incremental development of advanced human behaviour? No evidence of such ancient boats exists, although the long-term survival of perishable artefacts in tropical Asia would be highly unlikely; their possible existence will almost certainly never be proved either way. However, the fact that stegodons and other animals had presumably made the same journey to Flores by accident, as had countless other species that reached countless other islands without boats, went largely unnoticed in this excited discussion.

I was on a palaeontological expedition in the Egyptian desert in 2004 when I heard the first rumours that something very exciting had just been found in Indonesia. The dig was led by Elwyn Simons, one of the great figures of twentieth-century palaeontology and now reaching the final years of his career; it was a huge privilege to have been part of one of his last expeditions. As we gathered round him in the main fieldwork tent one evening, he told us all about news on the academic grapevine that a remarkable tiny new species of extinct human had just been discovered in a cave on Flores. It was anthropologically unprecedented. Elwyn was a natural raconteur, and his eyes sparkled as he told us what he knew of the discovery.

Previous fossil discoveries of ancient humans in Southeast Asia had been made at open-air sites, such as in river deposits at Trinil and Ngandong. Indeed, some older authorities considered that, whereas people sheltered in caves in Europe during the Ice Age, making it easy to find their prehistoric remains, caves in the tropics had been seen in the past as just 'the dwellings of bats, snakes and the great monitor lizards, and of course also of evil spirits.'[43] In the words of another authority, 'That is why excavations in caves in the tropics never bear any results.'[44]

However, excavations in Indonesian caves did start to yield some interesting archaeological discoveries. In the 1950s and 1960s Verhoeven had found Neolithic remains in several caves on Flores, made by modern humans during recent prehistory after they spread across Southeast Asia during Jared Diamond's 'Chinese steamroller' population expansion event. These included a large cave called Liang Bua (meaning 'cold cave' in the local Manggarai language). At the time, Liang Bua was being used as a makeshift elementary school classroom and was full of bamboo benches and tables, but once the classroom was moved, Verhoeven excavated a couple of metres down and found stone artefacts and pottery, and then six Neolithic burials with grave goods.[45] Further excavations continued periodically at Liang Bua over the following decades, and in 2001 an Australian-led team visited as part of a project to map the archaeological spread of modern humans across Wallacea to Australia. The deeper deposits at Liang Bua proved to be astonishingly rich in fossils – within a few years, the team had discovered at least 47 different stegodon individuals in the cave. And in September 2003, they made one of the most important and unexpected archaeological discoveries in history: some extremely fragile remains, including a fairly complete skull and jaw, right leg, partial pelvis and other fragments from a single female individual, which represented a completely new type of extinct human.

Much has been written about the 'hobbit', as the remarkable find from Liang Bua quickly became known. It was formally described as a new species called *Homo floresiensis* in late 2004, and what was truly shocking was its size. Its discoverers determined that although it was skeletally an adult – its tooth wear suggested an age of around 30 – its cranium had a volume of only 380 cubic centimetres. In contrast, modern humans have an average endocranial capacity of 1,350 cubic centimetres, and even extremely primitive hominins such as the African australopithecines, the first human ancestors to walk upright a few million years ago, typically had larger brains. The hobbit's brain was instead comparable in size to that of a small chimpanzee. The Liang Bua team estimated its height at 106cm (which they admitted might be an overestimate), and its mass at 36kg.[46] It was ridiculously, unfeasibly small, and completely unexpected. Peter Brown, one of the scientists involved with the original excavation, later said that he would have been less surprised if someone had uncovered an alien.[47]

Unsurprisingly, *Homo floresiensis* immediately became the focus of tremendous controversy, with many palaeontologists simply refusing to accept that it could be a real human species. The Liang Bua team considered that *Homo floresiensis* was descended from a population of another extinct hominin with more typical body proportions, probably *Homo erectus*, that had somehow migrated eastward across the Lombok Strait, been isolated on Flores, and became reduced in size in accordance with the island rule. However, as with the earlier debate around how archaic hominins could have reached the remote islands of Wallacea, many anthropologists either seemed unfamiliar with patterns of evolution shown by animals on islands, or refused to countenance the suggestion that insights from non-human species could apply to an extinct human. Dissenting voices maintained that

the hobbit's remains were instead those of a modern human with a
congenital defect or pathology, possibly microcephaly or Down's
syndrome; or maybe they were just bones from one of the Rampasasa
pygmies, who lived not far away on the island.[48] One group of critics
even suggested (erroneously) that the hobbit skull contained a dental
filling, 'evidence' that it must be a deformed skeleton from no earlier
than the twentieth century.[49] 'It's a complete circus,' said Brown at the
time.[50] These rejections of the hobbit's biological reality echoed
similar responses to the nineteenth-century discovery of Neanderthal
remains in Europe, which were dismissed as merely the bones of a
Russian Cossack with rickets dating from the Napoleonic War.
Ironically, Eugène Dubois was also convinced that Neanderthals were
just pathological modern humans, which prompted him to travel to
Java to eventually discover his *Pithecanthropus.*

How could a human with such a small brain have functioned, let
alone apparently exhibited complex social behaviours such as
harnessing fire and hunting large animals, as suggested by the
presence of stegodon bones with charring and cut marks in the Liang
Bua deposits? Was brain organisation, rather than size, the crucial
factor that allowed complex behaviour in hominins? Attempts to
understand the hobbit's evolutionary relationships through genetic
analysis proved unsuccessful due to rapid degradation of DNA under
moist, hot tropical conditions (a problem faced even in relatively
young tropical bones). However, further discoveries have vindicated
the Liang Bua team. The fragmentary remains of at least eight other
individuals were found in the cave, with additional hobbit bones
subsequently found elsewhere on Flores, demonstrating that the
skeletal proportions of *Homo floresiensis* did not match any known
human developmental abnormality or pathology and that the
remarkable characteristics of the first specimen were found across a
whole population.[51] More recently, the teeth, hand and foot bones
and femur of a different tiny hominin species have been found in the
Philippines and named *Homo luzonensis* in 2019. Although this second
tiny extinct human is very poorly known in comparison to *Homo
floresiensis,* its discovery indicates that evolutionary dwarfing seems to
have been a general phenomenon for hominin populations isolated
on islands.[52]

So when was the hobbit around? Whereas the Palaeolithic stone tools that first indicated the presence of an archaic hominin on Flores were up to 800,000 years old, the Liang Bua remains were initially thought to be only 18,000 years old, and maybe as young as 12,000 years old. These dates suggested not only a long evolutionary presence on Flores, but also lengthy overlap with modern humans, who reached Southeast Asia and Australia around 50,000 years ago.

Unique evidence has recently been found on nearby Sulawesi, showing that prehistoric modern humans soon exploited the naïve faunas they encountered on these remote tropical islands. Cave art depicting human–animal figures – 'therianthropes' – hunting Sulawesi's endemic warty pigs and dwarf buffalos has been dated to at least 43,900 years old.[53] A life-sized red ochre painting of a warty pig in another cave was recently confirmed to be older still, at least 45,500 years old.[54] In contrast, most of the famous cave art of Europe, such as the stunning images of wild horses and aurochsen on the walls of Lascaux Cave in France, is only between 21,000 and 14,000 years old.* These paintings are the oldest known representational art in the world; they have been described as 'the world's oldest recorded story'.[56] Intriguingly, however, both warty pigs and dwarf buffalos managed to avoid going extinct and still survive on Sulawesi today (although they are now threatened species). Might the hobbit therefore also have survived contact with modern humans?

More recently though, as with other sites such as Ngandong, dating of the youngest deposits at Liang Bua that contain hobbit bones and stone tools has been revised. The new dates suggest these deposits are actually no younger than 50,000 years old,[57] indicating a different prehistoric pattern – the disappearance of this enigmatic Caliban as soon as modern humans arrived on the island. This suggests a depressingly familiar conclusion. Should these remarkable island hominins therefore join the long list of naïve, defenceless island species – alongside the Little Swan Island hutia and Estanque Island deer

---

* Some simple European cave art is substantially older and may have been painted by Neanderthals. Ancient hand and foot impressions from the Tibetan Plateau, dating between 226,000 and 169,000 years old, have been suggested to be deliberate art made by Denisovans, another enigmatic extinct human species.[55]

mouse, the Madagascar sloth lemurs and weird Mediterranean cave goat *Myotragus* – that became extinct as soon as modern humans appeared on the scene?

However, we need to remember Darwin's idea of the fossil record as 'a history of the world imperfectly kept', where only a few scattered lines from a few scattered pages of the past diversity of life have been passed down to us. Liang Bua is just one site; just because rare hobbit fossils are only preserved in the oldest layers in this particular cave, this doesn't necessarily tell us much about what happened in other landscapes elsewhere across Flores. And it's important to recognise that gaps exist in the fossil record not only across space – affecting our relative understanding of past faunas on different islands – but also across time. There is no reason to expect the youngest known fossil of a now-extinct species to represent the last-ever individual of that species, or even to have existed at a similar time. Although the youngest specimens of abundantly-preserved and well-studied extinct species might correspond closely in age with their true extinction dates, this assumption is less likely to hold true for species with very few dated fossils – it's extremely unlikely that, by chance, a rare specimen just happens to date from the very end of a species' geological lifespan. The mismatch between a species' last known occurrence in the fossil record and its subsequent real (but probably unknowable) extinction date even has a formal name, the Signor-Lipps effect, named after the two American palaeontologists who first highlighted this conceptual problem; it's also known as the Romeo Error, after Romeo's mistaken assumption of Juliet's death that precipitates the tragic finale of Shakespeare's play. Indeed, numerous examples exist of so-called 'Lazarus taxa', species or evolutionary lineages that 'play dead' and vanish from the fossil record only to reappear millions of years later, and sometimes even turn up alive and well in the present – such as the 'living fossil' Bulmer's fruit bat, bush dog, Chacoan peccary and false killer whale that we've already encountered.

Evidence from an unexpected source has raised the tantalising suggestion that, just maybe, *Homo floresiensis* might have persisted until much more recently than 50,000 years ago. Gregory Forth, the anthropologist I'd met at London Zoo, has spent a huge amount of time since the mid 1980s conducting fieldwork in rural Flores with the Nage, an Indigenous group living in the central part of the island. Long before tiny hominin remains were discovered at Liang Bua, several middle-aged and elderly Nage informants told Forth similar stories about a very curious creature called the 'ebu gogo'.[58] This name possibly derives from the respectful term for an older lady (the same name we'd given to Ibu Jen), and the word for 'greedy' or 'gluttonous' – literally meaning something like 'glutton-granny' – although other derivations are possible, and 'ebu gogo' can also be used on Flores to refer to a type of spider. The ebu gogo were said to be a kind of hairy human-like being, smaller than the Nage themselves – maybe about a metre tall, although somewhat hard to estimate – that had lived in a cave called Liang Ula in the 'Ua region several generations earlier. They lacked any material culture or technology such as the ability to use fire or tools; they spoke in a strange, mumbling way and mimicked human speech parrot-fashion; and Forth's informants said they stank like the nests of the Flores giant rat *Papagomys*. They were said to be scared of dogs; and also, strangely, of combs (possibly because of their long hair). And as their name suggested, they were very greedy. Although the Nage had sometimes interacted with them, for example by giving them food when they appeared at feasts (where they would watch the human celebrations), the ebu gogo also regularly stole crops from gardens and fields, and this was to be their downfall. The Nage became so irritated by this persistent crop-raiding that they filled Liang Ula with 500 bales of palm fibre and set it alight, burning all the ebu gogo to death in their cave – although supposedly a pair who were out foraging managed to survive, and fled to another mountain.

Forth noted that although some informants reported fantastical details in their stories – for example, describing female ebu gogo as having breasts so long that they were carried over their shoulders, and could suckle their babies from behind – in general the Nage described the ebu gogo in naturalistic terms, similar to how they described 'real' animals known to still occur in the region. Indeed, they explicitly said that ebu gogo were not spirits or mythical creatures, but instead classified them somewhere between animals and people – possibly even as 'ancient humans'. Even the dramatic method that the Nage described their ancestors using to dispatch the ebu gogo in their cave turned out to be a method still used fairly regularly by people in central Flores to get rid of troublesome animals such as feral cats.[59] Could the tale of the ebu gogo therefore be more than just a story – might it instead be grounded in empirical reality? The discovery of *Homo floresiensis* suddenly provided a shocking possible candidate for the true identity of the supposedly fictional ebu gogo. The similarities were eerily close. Based upon the number of generations ago that the Nage said their ancestors killed the ebu gogo in the conflagration at Liang Ula, Forth estimated that, were this to have been a real event, it probably occurred sometime between 1785 and 1885.

Instead of dying out 50,000 years earlier, then, might the hobbit have actually survived until just a couple of centuries ago? Or was it even extinct at all? The Australian researchers involved with the discovery of *Homo floresiensis*, who had also heard stories about glutton-grannies during their previous decade of working on Flores, were impressed by their fossil's similarity to the tiny almost-people described by the Nage; some of them were open to the tantalising possibility that a few hobbits might conceivably still survive somewhere in the island's remote jungles. Even Henry Gee, the senior editor of *Nature*, was caught up in the excitement that mythical human-like creatures might be 'founded on grains of truth', announcing that 'Now, cryptozoology, the study of such fabulous creatures, can come in from the cold.'[60]

In addition to the ebu gogo, Forth described accounts of several other supposed types of wildman reported from Flores and elsewhere across Southeast Asia. He has recently documented another small hairy 'ape-man' described by the Lio ethnic group elsewhere on Flores, which was also allegedly wiped out by being burned in a cave – but sightings have been reported into recent years.[61,*] He also mentioned that he'd first been prompted to investigate wildman myths on Flores as a result of earlier stories he'd heard elsewhere in Indonesia, during his first major period of anthropological fieldwork.

---

* This entity supposedly jumps or hops about, is accused of stealing piglets, chickens and eggs, and – bizarrely – is sometimes reported as tickling people.

In the mid 1970s, Forth had lived in East Sumba for 22 months while conducting research for his doctorate at Oxford.[62] While there he learned about a quite different sort of wildman that local people told him still lived in forests and caves in the island's remote interior. It was very shy, generally hiding from humans and being extremely scared of dogs, and it supposedly liked to eat snails and slugs. Although some details inevitably differed between different stories, many of the regularly reported characteristics of the Sumbanese wildman were similar to those of the ebu gogo – it was a hairy human-like being with a strong animal-like odour, long breasts, and a limited ability to speak in a somewhat incomprehensible way. Unlike the ebu gogo, however, and more like the mysterious *Meganthropus*, it was instead usually described as very tall. This creature was given various names, but was most commonly called 'mili mongga'. The wildman featured in various Sumbanese folktales and myths, but intriguingly, Forth was also told several stories about people who claimed to have encountered mili monggas within living memory; unlike the ebu gogo, then, this creature still seemed to be part of the living landscape as perceived by local people in rural Sumba.

So, my slightly romantic speculations went as follows. Flores had until recently been home to its own remarkable unique species of (probably)-now-extinct endemic human, which, just possibly, might have been the inspiration for local legends of tiny people. But what about Sumba, the large island only 50km to the south, which had been largely ignored by palaeontologists – but which was washed by strong marine currents that flowed past Flores, and that could easily sweep Floresian waifs and strays over the sea and enable them to make landfall on a new island home? If stegodons had definitely made it to Sumba and evolved into a new endemic species, then what about the rest of the remarkable fauna documented in the recent fossil record of Flores? Might weird prehistoric humans have been washed out to sea and survived an overwater journey from Flores to Sumba, too? I liked Franz Weidenreich's advice from the 1940s about fieldwork in Indonesia:

> *The discovery of the Java man ... shifted the question of the missing link out of the stage of pure speculation into that of facts. The new discoveries solved*

*the Pithecanthropus puzzle but, at the same time, confronted us with new
and more specific problems. These, too, can be solved. The only requisites are
a spade, a hoe, and a little money ... The chances for rich rewards are much
greater here than they ever were in similar cases elsewhere.*[63]

Admittedly, Weidenreich also believed we had all evolved from
imaginary Southeast Asian giants, but his fieldwork advice still
seemed sensible. And anyway, Sumba seemed to have its very own
imaginary giant. Was there any chance that the mili mongga stories
collected by Forth might, just conceivably, have some connection
to a scientifically unknown human species that had once lived on
the island?

Back at the hotel in Waingapu, we cleaned ourselves up and changed
into fresh clothes for dinner. As we sat outside with a round of Bintang
beers, discussing the day's fossiling and the plan for tomorrow, Umbu
suddenly appeared again. He had an excited expression. 'You asked
about the mili mongga... I phoned some people I know, and I have
found out something more about it. My friend just told me that two
mili monggas came into a village in the countryside not that far from
Waingapu, and the villagers tricked them into eating corn with gravel
mixed into it. The mili monggas died, but their children were taken
in by the village. One of the children – the *anak mili mongga*, or child
of the mili mongga – is still alive, and it's being held in the village. We
could go there tomorrow. Do you want to go and see it?' We all stared
at Umbu. To say that this blew tomorrow's fossil-hunting plans out
the window would be an understatement. Day two of the Sumba trip:
the Holocene Dream Team discover a living wildman... Not quite
what we had expected, but it would do!

   After Umbu had left to organise our travel plans for the next day,
we talked excitedly late into the evening. It *couldn't* be what we were
all hoping, could it? There had to be another explanation, something
more prosaic ... but there didn't seem to be much room for
misinterpretation or miscommunication in the story that Umbu had
just told us; he had described an apparently recent event, and one that
was backed up with direct, living evidence. You couldn't ask for a
better situation to obtain concrete proof about a 'mystery man-beast'.

And yet... if it was really this simple to discover a live wildman on Sumba within a couple of days on the island, why hadn't anyone managed it before? It couldn't really be because only a handful of researchers had apparently ever been here... could it? Something felt like it couldn't be quite right with what we'd just heard; but the answers would have to wait until morning.

CHAPTER FOUR

STORYTELLING

But history is not acceptable until it is sifted for the truth. Sometimes this can never be reached.

Patrick White, Voss

It was hard for us all to sleep. Umbu arrived early, as we were finishing a quick breakfast washed down with strong gritty coffee. He seemed as fascinated as we were by the mystery; he'd never really thought much about the mili mongga before, and wanted to understand what it all meant too. We drove for a couple of hours along quiet dusty roads until we reached a small breeze-block farmhouse with a rusty corrugated roof, beside a field of dry brown grass being grazed by a couple of tiny Sumbanese ponies. A man in his sixties wearing an unbuttoned plaid shirt and wrapped in a sarong came out of the house and leant on the gate; Umbu went over to talk to him, and came back to tell us it was alright to go in. We introduced ourselves politely to the family inside and sat down. Umbu explained about our interests and began asking questions gently on our behalf. The man who'd met us outside nodded when Umbu mentioned the mili mongga. Yes, he said, it was true; the mili monggas had come to their village.

There were many taboos about discussing sacred and ritual things, it seemed; but some information about mili monggas could be shared with outsiders. They lived in the forest, and sometimes in caves. Nobody had seen one recently, but they were witnessed in the past. He described them – they were tall and had dark skin, black or brown in colour. They didn't wear clothes and were covered in hair, and the females had long, pendulous breasts.

He could also tell us about what had happened in the village. About five generations ago, a mili mongga had mated with a local woman and she'd had its baby. There hadn't been any stigma in having a mili mongga's baby, and the family carried on living in the village. The child had looked quite a lot like a mili mongga. The elderly man then pointed to his wife, a woman in her sixties sitting in the room with us; she is a direct descendant, he said. I turned. She didn't seem different from anyone else we'd met so far on Sumba. This wasn't quite the story relayed to us the previous evening. The old woman confirmed there wasn't any shame in being descended from a mili mongga, and you couldn't tell anything about her special ancestry by looking at her. But it had re-emerged in one of her relatives, she said. From her description, it seemed that saying someone was descended from a mili mongga might be a way to explain some kind of physical or mental disability, possibly offering a cultural context for understanding why such things occurred; but she made it clear that the relative was loved and supported by the local community. We drank coffee and chatted more about life in the village, and thanked the old couple profusely for taking the time to talk with us.

It was still reasonably early. We didn't turn back to Waingapu; Umbu told us he had some other contacts who might know more about mili monggas, so we continued inland.

The dirt road to Maoramba ran for miles along a narrow ridge, with forest dropping down steeply out of view on either side. When we eventually arrived, the village turned out to consist of a few huts huddled tightly against the sides of the road, partially hidden by huge green waxy leaves that drooped down from tropical trees growing along the ridge. It seemed very quiet. We got out of the car. Umbu was sure there was someone in Maoramba who could talk to us about mili monggas, but there was nobody to be seen.

I climbed up onto the deck of the biggest hut and stepped through the open door into the blackness inside. 'Hello?' I called. As my eyes adjusted to the darkness, I realised with a start that I was not alone. About 15 people were sitting cross-legged all around me inside the hut, staring silently. I froze. One of the silent villagers leaned forward and spat a stream of bright red saliva onto the floor. They were all chewing betel nut.

Betel probably doesn't match the requirements of what most people would think of as a 'fun' drug. It is both fiddly and actively unpleasant to take, and in my experience has little pay-off in terms of stimulation (although apparently the effects vary from person to person). The process of ingesting it is complicated and requires an entire kit, which is usually presented to you in a special little woven basket. The main ingredient is the betel nut itself, hard coin-sized slices of the dried seed of the areca palm, which is grown everywhere in local plantations across tropical Asia. After popping a couple of these into your mouth, you unscrew a little container of slaked lime (typically made from cooked and crushed seashells or coral) and tap some of the white powder into your palm, then dip a catkin of the betel plant into the powder (the betel plant is unrelated to the areca palm and actually a member of the pepper family). You put the powdery catkin into your mouth as well and chew everything up together, keeping it all in your mouth instead of swallowing. This combination of ingredients is called *sirih pinang* in Indonesian; the catkin is the *sirih* and the nut is the *pinang*.

The dusty-red betel nut is incredibly dry and powdery at first, sucking all the moisture from your mouth, whereas the texture of the catkin is like a stale damp carrot or a prawn; it gives a bit too easily when you bite into it. Then the moisture suddenly returns, and as the lime reacts with the nut you find your mouth filled with copious quantities of sour bright-red saliva, which you have to keep spitting out everywhere – real-life beetlejuice. By this stage, you begin to feel dazed. You then have to keep chewing and spitting until the desire to gag it all up overwhelms you, at which point you try to fight through the overpowering light-headedness and find a polite place to spit out the messy concoction, ideally not on the floor right in front of your host. On nearby Flores, people who just sit and chew betel nut as though they're chewing the cud are sometimes called 'buffalos in the shade'.[1] Betel nut is incomprehensibly popular from New Guinea to China; streets across tropical and subtropical Asia are stained with scarlet streaks of betel spit, and its use is banned in airports and hotels by cartoon posters showing people vomiting red fluid. The mouths and teeth of regular users are permanently stained red – on Flores they're said to have a mouth like a kingfisher's bill[2] – and such users are at significantly greater risk of developing a range of cancers. However, betel nut is hugely culturally important; different Asian cultures have each developed different combinations of chewing ingredients, and in rural Indonesia you will automatically be offered betel nut in the same way you'd be offered a cup of tea when visiting someone in the UK. In fact, I'm surprised that no one has opened a hipster betel nut café in Shoreditch.

I mumbled my apologies and stepped back into the sunlight. Back beside the car, Umbu had managed to find another local, who phoned someone else and told us to wait here – the person who could tell us about the mili mongga was on her way and would be here shortly. We all sat on the deck of the nearest hut and waited. A pair of pigs began mating vigorously right in front of us. It was very distracting, and carried on for longer than I expected. In order to give the pigs some privacy, Jen and I stretched our legs and walked down the road. We were gone for a few moments. As we walked back, Umbu greeted us with a wide grin, and we saw matching grins on the faces of Tim and James. The boys' mouths were bright red, and they looked completely dazed. Tim bent forward and spat a huge gob of red fluid all down his front. 'Oh for God's sake,' I laughed, exasperated. 'We leave you alone literally for a minute…!' Jen and I looked at each other and rolled our eyes. We were clearly the two sensible ones on this trip, we thought… as we waited for someone to turn up and tell us a story about the local yeti. 'Budge up and give us some room to sit down,' I said. Tim stared at me blankly and spat all over himself again. 'My mouth hurts,' he said.

A motorbike puttered towards us along the dusty road and pulled up in front of the hut. Its driver was an old lady, aged about 70, with a baby swaddled in a kind of papoose strapped tightly to her front. She got off the bike, reached under the baby's head for a packet of cigarettes and a lighter, lit a cigarette and tucked the lighter back under the baby. She took a deep drag and introduced herself. Umbu explained what she was saying. 'She asks if you want to know about the mili mongga? She says she knows about it because she's seen one.'

The old woman took another drag on her cigarette and told a story similar to the one that Umbu had given the previous evening, although now it was set further back in time. 'It happened at Miurumba village, in 1901. I heard from my grandmother that two mili monggas came into the kampung from a cave to look for food.

They couldn't speak. There were lots of them living in the forest then, but you can still find signs of the crabs and snails that they eat today, so they're still around. They had black hair, like a beard, all over their bodies – on their backs and on their arms – and their skin was black underneath their fur as well. The villagers gave them corn, but they mixed pebbles in with the corn, and the mili monggas died – they fell off a cliff near the village. But they had a child with them, and it was taken in by the village. It actually lived with my grandmother, in a kind of enclosure. Because it was brought up in the village, it was able to speak. It lived in the village for a long time, and only died in 2000.'

We definitely needed to start looking for fossils soon; but we were all captivated by the stories we'd just heard. I was reminded of a tantalising scientific paper written by the American palaeontologist David Burney and his Malagasy colleague Ramilisonina, who described a memorable research trip to Belo-sur-Mer in remote western Madagascar in 1995. Whilst trying to relocate fossil sites in the region that had been excavated in the nineteenth and early twentieth centuries, the scientists found that several villagers claimed to have seen or heard unusual large animals they called *kilopilopitsofy* and *kidoky*. These animals didn't match any of Madagascar's known living species, and descriptions given by the villagers sounded remarkably similar to the island's supposedly extinct dwarf hippos and giant lemurs (however, it's worth mentioning that one of the villagers, a local sorcerer, also told them about a mongoose that could allegedly kill rats and snakes by putting its tail down holes and farting). Burney and Ramilisonina were told that a woodcutter called André, who spent a lot of time in the forest, might be able to take them to find these mystery beasts … but, frustratingly, they were only able to meet André just before they had to leave – and were unable to arrange a trip into the forest with him.[3] So, *just* maybe, Madagascar's last dwarf hippos and giant lemurs are still hanging on in the remote forests near Belo-sur-Mer … but no scientists have yet returned to look for them. I didn't want the mili mongga to end up as another 'What if…?' like the *kilopilopitsofy* or the *kidoky*. We all decided that we had to follow this new lead.

Miurumba was much larger than the other villages we'd visited. Traditional Sumbanese houses, their peaked triangular thatch roofs stretching up into the sky, spread out around the large open area of dry trampled grey grass and earth where we parked. One of the tall roofs was decorated with buffalo skulls. Umbu chatted with the head of the village, a man in his forties who had come out to welcome us, and we were invited into his house for coffee, cigarettes and betel nut. There were holes specially cut into the raised rattan floor to allow you to spit betel nut saliva down onto the ground below the hut; but it was really difficult to aim the constant flow of red spit with any accuracy.

Sitting cross-legged on the floor, the headman told us the history of his village. It was named after a creature called the meu rumba, a word that meant 'wild cat', because it had supposedly once been visited by this being. I remembered that Gregory Forth's book gave 'meu rumba' as an alternative name for the mili mongga. But the meu rumba had been no conventional feline – it was like a lion but with a human head and legs, and it walked upright. The headman said that the female meu rumba had breasts like a woman, and meu rumbas breastfed and apparently lived 'like humans', although they didn't live in houses. However, the Indonesian word the headman used to describe the hair covering the meu rumba's body was *bulu*, which refers to animal fur rather than human hair. Unlike the people we'd spoken to earlier that day, he also said that the creature had a tail about 75cm long, differing from descriptions of the more hominin-like mili mongga.

Despite providing a different name and description for the creature, though, what had happened when the meu rumba visited Miurumba was similar to the stories we'd heard about mili monggas. The meu rumba had come into the village and entered the ceremonial building, and the villagers had given it corn to eat, which they mixed with pebbles that had been placed in a fire until they were red-hot. This had supposedly happened about a hundred years earlier, when the village was located in the hills closer to Waingapu (it had subsequently been moved because of a local war); although apparently people had seen similar creatures around here since the village had moved to its current location.

But what about the *anak mili mongga*, and the possibility of finding actual physical evidence of a mystery wildman – a possibility, however faint, that had brought us to this village where the stories we'd been told had apparently taken place? To our disappointment, the headman said nothing about a wild child that had been taken in by the village. However, he told us, there *was* a 'child of the mili mongga'... but of a different kind. At the site of the old village was a megalithic tomb with a carving of a mili mongga on it. There were supposed to be very large bones buried inside the tomb, and people prayed to it; apparently a 102-year-old woman still prayed there. The headman demonstrated the length of the bones by moving his right hand from the shoulder to the wrist of his left arm and back again, to indicate they were twice the size of a human arm; and then cupped both hands in front of him as though they were holding the sides of an imaginary chicken, as a measure of the thickness of the bones. During the nineteenth century, he said, a woman with no children had prayed at this tomb to become pregnant. When she conceived, she had called her child 'the child of the mili mongga'.

Had the meu rumba – or mili mongga – fallen off a cliff after being tricked into eating the burning pebbles? No, the headman said, but an hour's drive from here was a cliff containing four caves. One of them was called 'Gua Mili Mongga' – the cave of the mili mongga. One of the other caves was where the bodies of executed criminals were placed, another one was the place to put the bodies of people who have died in war, and the final cave was a place for rituals. An hour's drive didn't seem too far to go ... but the cliff became more and more inaccessible as we enquired further, now involving a long trek across rivers and requiring climbing equipment; and there were lots of dangerous snakes. It clearly wasn't somewhere we were going to be able to reach easily, or alive. We thanked the headman, finished our coffee and offered some final cigarettes to our host, and set off on the dusty drive back to Waingapu.

It had been a long day. The villages we'd visited had been several hours apart along bumpy country roads, and night fell as we were heading back. Whenever we stopped, everything outside was completely still; just the chirping of insects in the grass by the roadside and the warm tropical night air. We talked animatedly in the car about what we'd experienced since leaving Waingapu early that

morning. Well, *that* had all been different to what we'd expected...
and it was definitely a new experience for us all. I didn't have a clue
what was going on.

As well as being courteous and welcoming to some complete
strangers who turned up unannounced, everyone who'd spoken to
us had been serious about what they'd been saying. People from
several different isolated villages, remote from each other even by
car, had matter-of-factly recounted stories that all seemed to refer to
something that had happened a century or so ago. Did these accounts
refer to some genuine historical event that had been significant
enough to become part of local lore? They certainly hadn't felt like
folktales, the so-called 'stories from long ago', 'stories performed
with songs', 'stories that are held or attached', or 'voices of the
ancestors' that constitute ritual retellings of a heroic, mythical past
on Sumba.[4] And each story seemed to be a variation on what had
originally happened; was this because different communities had
differing levels of recall or understanding about the same events? As
we'd heard each new tale, it had felt like we were peeling back more
layers and getting closer to some core 'truth'. Details seemed to slot
suddenly into place, such as what the concept of the *anak mili mongga*
actually meant; this name had referred to a 'real' wildman child in
the first stories we'd heard, but then morphed into something that
seemed more plausible, and could have inspired garbled accounts of
adopted wild-children in villages. There was a spooky sense of
going round in circles but slowly starting to converge on something;
like finding pieces of a jigsaw that allowed us to reconstruct more
and more of a genuine historical event with each successive story
variant we were told.

But the stories were slippery and hard to pin down. The details
about children, cliffs, graves and bones still didn't quite align, and the
similar names whirled round in my head – Maoramba ... Miurumba
... mili mongga... All of these different takes on what might have
happened was like the plot of *Rashomon*... or rashomongga, I joked in
exhaustion. And if there had been an actual historical event behind it
all, whatever it was had happened just long enough ago that there
were no surviving witnesses, and no means of direct corroboration. It
was all deliciously frustrating. Despite all our discussion on the long

car journey, it felt like we were just scratching the surface. What were these stories all about?

Reports of mysterious wildmen are known from around the globe. Some of these beasts are world-famous and iconic, such as the sasquatch or bigfoot of America's Pacific Northwest and the yeti of the Himalayas, but there is a bewildering diversity of others out there too. I felt that I had a good background knowledge of these 'mystery man-beasts' or 'manimals'. As a child I had been captivated by an epic partwork magazine called *The Unexplained* and its volume on 'Creatures from Elsewhere'.[5] As well as containing riveting accounts of lake monsters and mysterious beings with names like Mothman and Owlman, its handy map of 'Man-beasts around the world' taught me about the South American maricoxi, the African chemosit, Australia's yowie, and the almas or almasty, chuchunaa, xueren and hibagon of continental Asia and Japan. Africa was also home to a man-beast that was excitingly referred to as just... 'X'. One of my primary school projects featured an enthusiastically reproduced version of this man-beast map complete with hand-drawn yetis and bigfoots (bigfeet?). Other books I picked up over the years referred to these creatures as 'BHMs' (short for Big Hairy Monsters) or in more sophisticated terms as 'cryptohominids', and described even more varieties – the kaptar of the Caucasus, the Sri Lankan nittaewo, the Florida skunk ape, and the barmanu of the mountainous borderlands between Pakistan and Afghanistan. Southeast Asia in particular seemed to be full of them – the most well-known was the orang pendek or 'short man' of Sumatra, but Gregory Forth's work also reported beings such as the lolok, To Ipono, To Uta and To Kaneke of Sulawesi, as well as the ebu gogo and various other seemingly distinct man-beasts from Flores.[6] The infamous Minnesota Iceman, a supposedly genuine apeman frozen in a block of ice and displayed as a fairground exhibit during the late 1960s, was also rumoured darkly at the time to be a BHM that had been killed in Vietnam.

Cryptozoologists – people who engage in the study of 'hidden animals' and investigate reports of mysterious, undescribed or 'out of place' creatures – have often taken a literalist approach to such accounts, interpreting these 'cryptids' as scientifically unknown flesh-and-blood creatures. From the 'father of cryptozoology' Bernard Heuvelmans onward, cryptids have regularly been interpreted as prehistoric survivors – the mokele-mbembe of the Congo is a late-surviving dinosaur, and the Loch Ness Monster is a remnant plesiosaur. Similar logic has been applied to reports of mystery man-beasts, as seen with the connection made rapidly between folk stories of the ebu gogo on Flores and the discovery of *Homo floresiensis* at Liang Bua. Cryptozoologists have also endeavoured to make stories of the yeti more credible by linking them with *Gigantopithecus*, a giant ape known from the Pleistocene fossil record of India and China. *Gigantopithecus* was another of Ralph von Koenigswald's important palaeontological discoveries, and one made in an unexpected location; it was described on the basis of some unusual large teeth that von Koenigswald found in a drugstore in Hong Kong in the 1930s, among fossil bones imported from southern China that were used as remedies in traditional Chinese medicine. *Gigantopithecus* was distantly related to orangutans, but is still only known from fossil teeth and jaws, making it very hard to reconstruct what it looked like as a living animal. However, it was definitely much larger than a gorilla and possibly up to 3m tall – a real-life 'Hong Kong King Kong', and one that evolved in a continental ecosystem where (unlike on islands) primates really could sometimes become giants.

Writing in the 1950s, Heuvelmans imagined a dramatic narrative for the evolution of the yeti from *Gigantopithecus*. These 'stupid or too peaceable great brutes', facing a constant onslaught from the prehistoric humans who were spreading across Asia, became slowly pushed up into the remote mountains of the high Himalaya where their last remnant populations could remain hidden. Heuvelmans admitted that his theory was 'utterly hypothetical', but he was convinced that it 'provides the only entirely acceptable explanation of the mystery of the abominable snowman.'[7] The idea that the yeti might be a late-surviving *Gigantopithecus* was even endorsed by some anthropologists, such as Carleton Coon, who considered 'that man may not be the only erect bipedal primate to have survived the Pleistocene period.'[8] Heuvelmans even proposed that the yeti should be given the scientific name *Dinanthropoides nivalis* (a literal translation of 'abominable snowman' into a mixture of Greek and Latin), but getting ahead of himself slightly, wrote that 'If one day its teeth are examined and found to be identical with those of the Gigantopithecus, its name will have to be changed, according to the rule of priority, to *Gigantopithecus nivalis*, the present species being no doubt quite distinct from the Pleistocene primate from Kwangsi'.[9],*

Another cryptozoologically-minded anthropologist, Grover Krantz, drew even closer parallels between *Gigantopithecus* and bigfoot, arguing that this mystery man-beast was in fact the same species as the supposedly extinct Chinese *Gigantopithecus blacki*. When this suggestion was dismissed as impossible to validate, Krantz suggested naming a new *Gigantopithecus* species, *G. canadensis*, on the basis of plaster casts of alleged bigfoot footprints (although he also suggested that bigfoot might alternately represent an unknown American species of the extinct African hominin *Australopithecus*).[10] A dedicated anthropologist, Krantz's skeleton (holding the skeleton of his wolfhound Clyde) is now on public display in the Smithsonian's National Museum of Natural History in Washington, D.C., to educate visitors about human anatomy. Ironically, the absence of comparable skeletal material for *Gigantopithecus* from the Asian fossil record gave cryptozoologists such as Krantz free rein to 'reconstruct' this enigmatic giant ape as a biped that closely resembled modern wildman accounts.

Heuvelmans and other cryptozoologists considered that many man-beasts are descended from other 'officially' extinct hominins or fossil primates. North American man-beasts are supposedly the direct descendants of dryopithecines, an extinct group of apes that lived in Eurasia and Africa during the Miocene and Pliocene epochs, and the Minnesota Iceman and central Asian man-beasts might be late-surviving Neanderthals.[11] Two Russian cryptozoologists, Dmitri Bayanov and Igor Bourtsev, wrote in the academic journal *Current Anthropology* in 1976 that 'the hominologist's dream is that all the hominid forms known from the fossil record, and even those not known from it, will turn out to be alive'.[12] The 'prehistoric

* 'Kwangsi' is the old transliteration of Guangxi Province, China.

survivor' model has given stories about mystery man-beasts a gloss of apparent scientific plausibility, and from Stig of the Dump to the cavemen racers with their Boulder Mobile in *Wacky Races*, the idea of prehistoric humans surviving into the present has obvious popular appeal. While conducting fieldwork in the Caribbean a few years ago I found a dog-eared paperback in the research station I was staying in, with the gold-embossed title *Neanderthal: Their Time Has Come* – about a pair of Harvard palaeontologists (and former lovers) who follow up tales about yetis to find a relict tribe of telepathic Neanderthals in the mountains of Tajikistan. It was the perfect book to read under a mosquito net by the light of a head-torch.

There are, of course, huge problems with the idea that reports of mystery man-beasts represent prehistoric survivors. New species of primates, including apes, continue to be described today. I myself have helped to describe a new species of living gibbon from China, the endangered skywalker hoolock gibbon, and another species of now-extinct gibbon found in the tomb of the grandmother of the first Emperor of China as part of a royal menagerie. However, nobody has provided any robust evidence to support the existence of an unknown living giant bipedal ape or secretive hominin. Such 'evidence' as is available consists almost entirely of unsubstantiated sightings, and all the physical evidence put forward over the years has been found severely wanting – from plaster casts of dodgy footprints and old photos of mysterious tracks in the snow, to frozen icemen in fairground attractions that mysteriously vanish upon investigation, to shaky film footage that people confessed to faking decades later. An alleged yeti finger smuggled out of India in the 1950s by Hollywood movie star James Stewart (hidden amongst his wife Gloria's lingerie) turned out to be from a historically deceased human, and a high-profile genetic study of hair samples reportedly obtained from bigfoot, yeti, orang pendek and the Russian almasty found they were all from known mammal species.[13] Although this study suggested that two 'yeti' samples were genetically closest to an extinct lineage of ancient polar bear, subsequent work showed these samples were just from the rare (but not mythical) Himalayan brown bear.[14]

As palaeontologists ever since Darwin have recognised, the fossil record provides a hugely incomplete account of the history of life on Earth. Cryptozoological suggestions of prehistoric survivors are therefore theoretically conceivable if we think in terms of Lazarus taxa and the Signor-Lipps effect. Might all these mystery man-beasts therefore represent yet more examples of living fossils, like Bulmer's fruit bat and the thick-toothed grampus? Well... no. It's just about possible that one of these supposed cryptids could conceivably be a late-surviving prehistoric human – most likely from a remote region known to contain geologically young hominin fossils, where relatively few fossil sites are known (so recent survival cannot be ruled out), and where thorough systematic surveys of existing biodiversity have not been conducted. The ebu gogo of Flores could potentially fit the bill as a contender... or maybe a cryptid from a neighbouring island like Sumba? However, the problem with mystery man-beasts is that they have been reported almost *everywhere*. Is it really plausible that *all* these critters represent scientifically undescribed bipedal primates that somehow conveniently failed to be preserved in recent fossil deposits, in addition to conveniently not leaving any indisputable evidence of their existence today?

Neanderthals and *Gigantopithecus* are known from numerous Eurasian Quaternary fossil sites. The most recent Neanderthal fossils date from about 40,000 years ago (in a now-familiar pattern, they disappeared around the same time that modern humans arrived in Europe), and the youngest dated fossil beds containing *Gigantopithecus* are over 400,000 years old.[15] The regions where these fossils occur have well-studied recent fossil records, with many sites that postdate the last known appearance of either species. Although numerous examples of Lazarus taxa exist, these are still very much the exception rather than the rule. The question is, then, why wouldn't late-surviving hominins have been fossilised if they were still around in the Late Pleistocene or Holocene?

The lack of such geological continuity of man-beast fossils into the past few tens of thousands of years – anywhere in the world – stretches the idea of fossil incompleteness past its breaking point. Even more problematically, many man-beasts are reported from biogeographic regions such as the Americas and Australia that lacked hominins – or apes of any kind – before modern humans spread round the globe a few tens of thousands of years ago. In the words of Darren Naish, a palaeontologist who provides thoughtful critiques on cryptozoology, 'the concept that a group of long-extinct animals might not only have survived to the present but might also occur in a part of the world where they have no record at all (fossil or modern) is too speculative to be taken seriously'.[16] This is the wider problem with the 'prehistoric relict' explanation for cryptozoological critters – whilst it combines the romantic appeal of a 'last survivor' with a semblance of scientific robustness, it's unfortunately more credulous than critical. Following this logic, sightings of something mysterious out at sea are explained away as prehistoric champsosaurs, and local accounts of half-glimpsed animals deep in a tropical forest must be long-extinct chalicotheres, regardless of the lack of any evidence for the occurrence of such obscure beasts in the fossil record for many millions of years.

The stories we'd been told about mili monggas and meu rumbas were unexpected and fascinating, seeming to hint that something more complex might be going on to explain them. It would be a huge shame – and potentially a missed opportunity – not to explore them further and see where they might lead. But did I really want to start looking into local reports of mystery man-beasts? Surely cryptozoology, and the credulous acceptance of highly implausible animals that it usually seemed to involve, would be the kiss of death for the career of a 'real' scientist... For example, Grover Krantz was seen as an embarrassment to the anthropology department where he worked, in the words of his relative Laura Krantz.[17]

However, there seemed to be something in the air in Southeast Asia that gave cryptozoology a bit more leeway in terms of respectability – maybe because this part of the world had such incredible amounts of tropical biodiversity, but so much of it was still poorly known? As well as conducting the only study on Sumba's mammal fauna in the 1920s, Karel Dammerman also collected and wrote about numerous colonial-era accounts of Sumatra's orang pendek.[18] Cautious acknowledgement of the possibility that the orang pendek or other Southeast Asian wildmen might really exist has been provided more recently by respected living zoologists such as John MacKinnon and Ian Redmond, and evidence to support the creature's existence has been exhaustively pursued by Debbie Martyr, who has also worked tirelessly for Fauna & Flora International over the past two decades to conserve the critically endangered Sumatran tiger.[19]

Even former scientists from my own institution, the Zoological Society of London, made public pronouncements about the possible reality of this particular wildman. Professor William Osman Hill, one of the leading twentieth-century authorities on primate anatomy and the Society's prosector (anatomical dissection preparator) for 12 years, suggested in 1945 that some orang pendek reports seemed very similar to what was known about the anatomy of *Homo erectus*. He made the remarkably prescient suggestion that both this wildman and the tiny nittaewo of Sri Lanka might be dwarfed island populations of *Homo erectus* that had managed to survive into the present, foreshadowing the discoveries of *Homo floresiensis* and *Homo luzonensis* by over half a century.[20] Osman Hill had also been the first scientist to look at James Stewart's smuggled 'yeti finger', and had further concluded that a supposed yeti scalp obtained by Sir Edmund Hillary was probably made from the skin of a goat-like ungulate called a serow (although he disproved the alternate hypothesis that it was made from a yak scrotum).[21] However, he remained open to the possibility that the yeti too was an undescribed flesh-and-blood animal, stating in 1961 in the wildlife journal *Oryx* that it was likely to be 'a plantigrade mammal capable of bipedal progression and of jumping', which inhabited the rhododendron forests of the Himalayan valleys.[22] However, in the same paper he dismissed 'an enormous faecal sample', supposedly from a bigfoot, that he had been sent from California – which 'is quite another story and need not detain us here'.[23]

But the stories we'd just been told about mili monggas and meu rumbas didn't seem that similar to the accounts of yetis, sasquatches or other mystery man-beasts I'd read as a child. Instead of tales from startled hunters or explorers who thought they'd encountered *something* in the woods, these stories all described some sort of event from long ago, and seemed to be layered with other meanings or allusions that remained unclear to me. The link between wildmen and graves was a curiously regional characteristic, too; Gregory Forth also mentioned a burial mound on Flores that supposedly contained a mass grave of hominoid creatures called *ngiung* … but that had proved impossible to investigate further due to local restrictions.[24] And the meu rumba apparently wasn't even a wildman at all, but instead some sort of weird bipedal lion that sounded even less plausible than a prehistoric hominin survivor. In fact, if the mili mongga hadn't been included in Forth's book, I wouldn't have made an automatic link to bigfoot or the yeti at all. What was actually behind these stories, then?

The long dusty road finally emerged from the hills and joined the main tarmacked coast road that led back to Waingapu. It was very late. Further consideration about the stories we'd heard would have to wait; it was time to find some fossils.

CHAPTER FIVE

RODENTS OF UNUSUAL SIZE

> *... and Antiquities are history defaced, or some remnants of history which have casually escaped the shipwreck of time.*
>
> Francis Bacon, *Advancement of Learning*

It was another long drive to Mahaniwa. We headed south from Waingapu into the rugged interior of the island, climbing slowly up a bumpy unpaved mountain road. The views were breathtaking; the land fell steeply away into a series of grey hills and gorges that stretched on for miles, under an expanse of sharp blue sky dotted with a few bright clouds. The unforgiving sun rose higher as the hours passed. There was little sign of life in this landscape of bleached grass and rock, which was disturbed only by shadows of clouds moving silently across the hillsides; but every now and again in the middle of nowhere we would drive past a spectacular megalithic tomb, like the stone table on which Aslan was sacrificed. The setting was certainly as otherworldly as Narnia. A profound sense of timelessness, of a place bewitched and becalmed, was also experienced by anthropologist Janet Hoskins during her stay on Sumba:

> *What I felt ... most vividly was not that I was confronted with an anachronism, that these people represented some archaic or obsolete way of life, but that I myself had come unstuck from any familiar temporal framework and was quite simply misplaced in time. Sumba seemed not a survivor from the past, but an area that was profoundly out of sync with the modern world, moving to its own rhythms ...*[1]

It was mid afternoon when we arrived at a handful of low houses with rusting corrugated iron roofs, dotted along the base of a long limestone cliff. We were a few thousand feet above sea level at the edge of East Sumba National Park, one of the few patches of tropical forest left on the island. A small river lined with lush green vegetation flowed out from the forest and ran along the side of the road. A hugely pregnant pig lay in the sun against one of the houses. Nearby, an ancient woman with wispy white hair crouched low against the ground. She was achingly thin, and was flattened down onto herself like a folded chair, long elbows sticking out at her sides. Slowly, laboriously, she was pounding coffee beans on a flat stone in front of her ... *tok tok* ... *tok tok*. The sound was hypnotic and sleepy, like the heartbeat of the village.

Umbu had brought us to Mahaniwa because the limestone cliff that separated the village from the forest was full of caves, and – we hoped – limestone caves might mean fossils. We settled into a daily routine of exploring the huge cliff face for possible fossil sites. The days were long and dusty; and a straggle of men and teenage boys from the village would usually accompany us up the slope as we poked around in the many caves along the ridge, smiling with good humour at our exploits. James always managed to bang his head on the low limestone ceilings, filling the caves with exasperated cursing.

Although we were high in the hills, this was still the tropics; the sun set suddenly at six, so we had to be back in the village before then. In the evenings we would have a small meal of boiled rice mixed with instant noodles. Somewhere really far up the road was a tiny shack that sold bottles of warm beer, so every few days one of us would make the journey on the back of a local guy's motorbike (usually Tim or James, because they were braver than me). Jen amused us by always making the effort to dress for dinner, throwing a black shawl round her shoulders; the rest of us were too mucky and exhausted. The power would go out shortly after sunset, so we went to bed early. Outside the hut, the night sky was deep black and dripped with brilliant stars. The Milky Way was splashed vividly across the heavens, full of other worlds that might even be as strange as this one.

We hit pay dirt in a cave called Liang Lawuala, a tiny opening halfway up the cliff face that could be reached by scrambling up a steep path of scree behind one of the houses along the road (made even more fun by having to avoid a pair of angry dogs at the side of the house). The cave mouth opened into a small sheltered 'porch' where we could all just about squeeze in and sit down, and with a narrow hole in the back wall. Crouching down uncomfortably, we could clamber through this hole into a claustrophobic passage of rock and compacted mud that descended steeply into the limestone cliff, which we had to Cossack dance our way down into the gloom. Several metres down, in complete blackness if we turned off our head-torches, a low tunnel curved backwards off the left side of the main passage. We could crawl along this tunnel with our hands and knees in the damp cave filth.

The tunnel's low curved walls and roof were covered with the strange mossy nests of cave swiftlets, glued on with saliva and looking like magical little fairy seats. These birds were close relatives of the famous Asian edible-nest swiftlets, which construct nests of solid saliva that fetch huge prices to make soup with supposed aphrodisiac and medicinal properties. If I shone my head-torch on these weird little nests, they seemed to glow a numinous green. And crawling silently and stealthily across the walls were amblypygids, huge tail-less whip scorpions with sickening flattened bodies, large spiny grabbing mouthparts, and grotesquely thin and elongated legs, which they used to blindly feel out their prey in the darkness. The narrow circle of light cast by my head-torch would suddenly reveal one of these things pressed tight against the cave wall right in front of my face, or just about to slide its long, thin thread-like grey legs across the back of my hand. I value and respect the intrinsic right of all biodiversity to exist, and the amblypygids were remarkable and evolutionarily fascinating; it was a privilege to see such an unusual life-form; but I really, *really* wanted them to not be all around me in this tiny low underground tunnel. I watched one creep into an empty swiftlet nest and hide there, and felt a bit dizzy. Maybe most unsavoury of all, halfway along this black tunnel metres and metres underground was a soiled damp sheet stretched out across the cave floor. Somebody, or some people, had clearly been down here – to what was undoubtedly the most unrelaxing and unerotic location I could imagine – and had presumably been lying on the sheet for... some reason.

And of course, it was here that we found fossils. Hunched down and crawling forward on our elbows, our faces a few centimetres above the cave floor, we started to see bones scattered over the clammy sediment. Some of these were the tiny delicate bones of

previous generations of cave swiftlets, but amongst these were other
rust-coloured or darker brown bones that belonged to something
else. We started to collect the bones systematically and store them in
sealable plastic bags. Gently scraping away with our trowels revealed
more and more old bones everywhere in the surface layers of sediment.
The air underground was close and musty, and weirdly hot; my shirt
stuck to my back with sweat and we were all soon covered in muck.
We crouched in silence, working away for hours. I shifted constantly
from one uncomfortable position to another to fight the pins and
needles in my legs and the sharp rocks digging into my knees. Every
now and again, one of us would glance uneasily back at the soiled
sheet we'd had to crawl past; after a while we became half-convinced
that, every time we looked, it had moved slightly closer towards us
along the cave floor.

There was no sense of time in this still, silent underground world.
So it was a surprise to hear a kind of whooshing sound coming from
the main passageway outside our tunnel. Surprise rapidly turned into
shock and panic – all of a sudden the air was filled with beating,
flapping *things*, moving incredibly fast all around us and swerving
horribly close about our heads. I thrust my arms up instinctively to
protect my face. I had no idea what was happening and it was
horrendous. Then I looked up – the things were darting in and out
of the mossy nests on the walls and ceiling. It was the cave swiftlets;
it must be almost sunset outside, and they'd come back in a flock to
roost. 'Turn your head-torches off!' I yelled – the birds were confused
by the lights and were flying straight at them, and at our faces. I
heard Tim start to yell a reply, but he only managed a weird muffled
coughing noise. He spat noisily. 'One of them just flew straight into
my mouth!' Our lights out, we groped for our bags of fossils and
crawled slowly along the tunnel in complete blackness, calling out to
each other to try to find our way, while the air above us and around
us was full of feathers and wings and clattering. It was like a terrible
dream. My head was pressed down right against the damp floor to
avoid banging it on the low ceiling, and I forced myself to breathe
slowly and calmly. I prayed that I wasn't going to suddenly feel the
soiled sheet touching my hand as I inched forward. Eventually I
found the muddy main passage and crawled upwards, following

somewhere behind Jen's voice. One by one Umbu helped us all to squeeze out through the narrow hole at the back of the entrance porch, to be greeted with sweet late afternoon sunlight and a panoramic view over the huts of Mahaniwa. We laughed in shock. James checked his watch; it was just after five o'clock. 'New rule,' I said. 'We *have* to keep an eye on the time, and be out of this cave before five.' Nobody disagreed. Back in the UK, I learned that the cave swiftlets of Sumba are genetically distinct from populations on nearby Flores and Sumbawa, and might be an undescribed species.[2] Tim had nearly swallowed a possible new species of bird – not many people can make that claim to fame.

The next morning we sat outside the headman's hut in the bright sunshine with our cups of strong black coffee and carefully sorted through the bones from the cave, our prize specimens laid out across a couple of white plastic chairs. We'd collected a haul of rat bones: mandibles, skull fragments and limb elements. Peering at the teeth within the jaws through a hand lens, I saw straight away that these weren't any of the invasive rat species found on Sumba today – black, brown, and Pacific rats, which had been accidentally spread around the world by people and were documented by Dammerman in his 1928 survey of the island's 'extremely poor' living mammal fauna. There seemed to be several types of rats amongst the bones we'd collected, which varied in both size and morphology. The largest mandibles were the most unusual and distinctive, with huge fluted molars bulging out from their jaws. The leg bones fell into a range of corresponding sizes, with the largest looking like they belonged to an animal as big as a ferret or mongoose – a truly giant rat, or *tikus besar* as we jokingly called it in Indonesian. Nothing like this had been reported from Sumba before. 'They're all something new!' I grinned to the others. Umbu peered carefully over my shoulder as I showed him the best specimens. We laughed and slapped each other on the back. These were spectacular fossils, and just what we'd dreamed of – thanks to Umbu and his hunch about these caves, we had found a new island mammal fauna of giant rats!

Rodents are fascinating and tremendously successful mammals. Although they often receive little attention compared to larger, more charismatic mammals such as carnivores, ungulates, whales and dolphins, they are remarkably diverse in ecology and morphology, performing vital roles such as seed dispersal in terrestrial ecosystems and representing an important part of food webs for vertebrate predators. They are hugely species-rich, with over 2,200 species known so far from around the world – more than a third of the 6,500 or so mammals – and with new species constantly being described. Only bats come close, with over 1,200 known species. All rodents possess a pair of continuously growing sharp incisor teeth, which allow them to gnaw and burrow and provide the secret of the group's success. I'm one of the co-ordinators of the International Union for Conservation of Nature's Small Mammal Specialist Group, which provides attention for the world's rodents and other neglected small mammal species – supporting their survival in a changing world is of vital importance.

Rodents also seem peculiarly good at getting onto islands, where they become genetically isolated from their ancestral mainland populations and are exposed to the different evolutionary selection pressures in island ecosystems. Possibly this is because continental rodents are typically small-bodied, fecund, and often arboreal, so have a better chance of being washed out to sea on a floating raft of vegetation or fallen tree, surviving the voyage, and being able to reproduce rapidly upon arrival. Many island groups – such as the Lesser Antilles, the Solomon Islands, and the islands of the Gulf of California – have native land mammal faunas composed only of rodents and bats (which are able to cross water barriers much more easily by flight). Rodents are also the only placental land mammals other than bats that made it all the way to the 'island continent' of Australia before humans. In addition to its well-known marsupials and monotremes, at the time of European arrival Australia had over 60 native rodents, including such species as the Darling Downs hopping mouse,* the Kakadu pebble-mound mouse, and the more humbly-named little native mouse, the end result of multiple colonisations and endemic evolutionary radiations.

Australian rodents are all 'regular' in size, but other island rodents became giants in accordance with the island rule. The largest of these

---

* Sadly the Darling Downs hopping mouse is now extinct.

was a beast called *Amblyrhiza*, known from fossils from the tiny Caribbean islands of Anguilla, St Martin and St Barts, which lived during an interval of Quaternary low sea level when these landmasses were connected together into a larger Caribbean palaeo-island. Body mass estimates for *Amblyrhiza* vary, but it may have weighed as much as 200kg, comparable to a brown bear.[3] It's likely that *Amblyrhiza* evolved from small mainland South American rodents called spiny rats, which weigh about 300g – if so, this is the largest size increase of any island mammal, living or extinct.[4] *Amblyrhiza* is one of the few extinct mammals known from the Caribbean Quaternary fossil record that almost certainly died out before humans reached these islands; instead it appears that rapidly rising sea levels associated with Ice Age glaciation cycles made its island home shrink and fragment, until the landmass was simply too small to support a viable population of these monster rodents.[5]

Other island rodents also display unusual patterns of convergent evolution, filling the niches of other mammals that are missing from depauperate island ecosystems. They have even become carnivorous on four independent occasions in the Philippines, Wallacea and Sahul (Australia plus New Guinea).[6] Some of these animals are really weird indeed – especially the shrew rats of Sulawesi, which, as their name suggests, are convergent on predatory shrews. This group of misfits includes a giant hog-nosed species almost half a metre in length called *Hyorhinomys stuempkei* or the 'Sulawesi snouter', which also has extremely long urogenital hairs – leading to news coverage in 2015 of 'Giant carnivorous rat with long pubic hair discovered in Indonesia'.[7] Its discoverers considered it so weird that they named it after the rhinogrades or snouters, imaginary mammals with hugely modified noses from the islands of 'Hy-yi-yi' that were dreamt up in the 1950s by German zoologist Gerolf Steiner (writing under the pseudonym Harald Stümpke);[8] reality here is at least as strange as fiction. Another Sulawesi shrew rat, *Paucidentomys vermidax*, was discovered in 2012 and is the only rodent with no molars; it apparently feeds exclusively on earthworms, which it presumably sucks up like spaghetti with its elongated bandicoot-like snout.[9]

Not all rodents perform useful services to biodiversity, though. Unfortunately some of the world's most destructive invasive species are rodents – the rats and mice that unwittingly accompanied explorers and sailors as stowaways onboard their ocean-going vessels, and managed to colonise even the most remote islands. Four species have proven to be particularly problematic at a global scale: black, brown, and Pacific rats, and also the seemingly humble house mouse, which all belong to the family Muridae. These species are all now distributed so widely that in some cases it has been challenging to even determine their native geographic ranges. Intriguingly, the Pacific rat – an invasive rodent adapted to tropical environments, that's found today across much of continental and insular Southeast Asia and far across the remote islands of Polynesia – might have originated on Flores, the next island to Sumba, based on patterns of genetic variation shown by different populations.[10]

The endemic birds, reptiles, invertebrates and plants that have evolved on islands for millions of years often don't stand a chance when the rats arrive. These invasive rodents are generalist omnivores and consume pretty much anything – allowing them to persist at high densities even as particular native species decline, because there are still other things for them to eat. And once rats have got onto an island, they are very difficult to get rid of. Although eradication programmes using poison bait have been conducted successfully on relatively small islands – up to the size of the subantarctic island of South Georgia, about 170km long – these programmes are logistically complex and hugely costly, and particularly difficult to conduct on islands with native land mammals at risk of taking the same bait.

It's thought that the disappearance of possibly thousands of endemic Pacific island bird species over the past 2,000 years was caused mainly by the accidental introduction of Pacific rats by Polynesian colonists. More recently, historical introductions of black and brown rats and house mice (all originally natives of eastern or Southern Asia) have driven further ongoing massive declines of island biodiversity. House mice accidentally introduced in the nineteenth century to Gough Island in the southern Atlantic have become 50 per cent larger than average and chew their way through almost two million seabird chicks and eggs a year – including those of the critically endangered Tristan albatross, which has 99 per cent of its world population on Gough Island.[11] And adversely affected islands aren't all tropical or remote;

the arrival of black rats to the British Isles during the Roman period around 2,000 years ago is associated with the disappearance of seabird colonies and extinction of Scottish gadfly petrels, which possibly represented a unique North Atlantic seabird species now lost forever.[12]

Many island endemics that survived this invasive rodent onslaught only managed to cling on as incredibly vulnerable remnant populations on tiny offshore islets; it was often nothing short of a miracle that these species didn't also become extinct. Campbell Island, one of New Zealand's subantarctic islands, was colonised by brown rats around 1828, and when naturalists first visited in 1840 they commented on the surprising absence of land birds. In 1997 it was found that a previously unknown endemic snipe, a cryptic small brown wading bird, had managed to survive unnoticed for over 150 years on rat-free Jacquemart Island, a sheer-sided basalt plateau a kilometre offshore and only 500m by 750m in area.[13] A similar incredible story of survival against the odds is that of the Lord Howe Island giant stick insect or 'land lobster', a huge flightless insect once abundant on Lord Howe Island between Australia and New Zealand. Black rats got ashore from a shipwreck in 1918, and the land lobsters were thought to have been eaten rapidly to extinction within a couple of years. However, in 2001 the species was rediscovered alive on Ball's Pyramid, an incredibly remote sea stack 20km from Lord Howe Island, with a total known population of only 24 individuals found living under a single shrub. Thankfully, both of these stories have happy endings – rats were eradicated from Campbell Island in 2001, and within a few years snipe had begun to recolonise naturally by flying over from Jacquemart;[14] and the land lobster has become the focus of a successful captive breeding programme involving multiple zoos (and with a few other wild individuals subsequently seen further up Ball's Pyramid in 2014).

Sometimes the arrival of rodents on islands has precipitated wider ecosystem collapse as well as 'just' extinction of native species. When Easter Island was first encountered by Europeans on Easter Sunday 1722, it was almost completely deforested. The archaeological record reveals the island was once covered in trees, including some of the world's largest palms (possibly a now-extinct endemic genus).[15] Historical records suggest some patches of woodland persisted into the nineteenth century, but today the barren denuded island is 'almost a moonscape in appearance', in the words of archaeologists Terry Hunt and Carl Lipo.[16] The loss of Easter Island's forests is often seen as the tragic result of clearance by Polynesian settlers who arrived by about AD 1200, a suggestion first made by French explorer Lapérouse in the eighteenth century.[17] Jared Diamond described Easter Island as 'the clearest example of a society that destroyed itself by overexploiting its own resources',[18] and promoted a hypothesis of 'ecocide' – with Easter Island held up as a salutary lesson of what might befall humanity more widely if we continue our oblivious destruction of nature. In Diamond's words, 'I have often asked myself, "What did the Easter Islander who cut down the last palm tree say while he was doing it?" Like modern loggers, did he shout "Jobs, not trees!"? Or: "Technology will solve our problems, never fear, we'll find a substitute for wood"? Or: "We don't have proof that there aren't palms somewhere else on Easter, we need more research, your proposed ban on logging is premature and driven by fear-mongering"?'[19]

However, the Polynesian settlers brought Pacific rats with them, and caves on the island contain many nuts of the now-extinct palm that show telltale evidence of rat gnawing. Palm nuts and seedlings constitute prime rat food, and if rats ate their way through the next generation of trees, the forests would have become stands of slowly ageing 'living dead' with no natural population recruitment.[20] This hypothesis is supported by evidence from other Pacific islands, such as O'ahu in the Hawaiian archipelago. Well-studied prehistoric sites show that although Polynesians and rats arrived on O'ahu together, rats spread across the island more rapidly; and dating of palm pollen in lake sediments shows that large-scale decline of palm forest matches the timing of local rat arrival rather than later expansion of human settlements.[21] Easter Island can thus still provide a salutary lesson – but one about the insidious effect of introduced species and the unwitting consequences that our actions have on biodiversity.*

---

* It is important to note that the main factor responsible for the collapse of Easter Island society – either the loss of forest resources, or subsequent European social disruption, which included slavery and the introduction of smallpox – remains the subject of debate and has significant implications.

And destruction to island ecosystems caused by invasive rodents isn't restricted to the terrestrial realm. Seabirds form large colonies on many offshore islands, and play a crucial role in transferring nutrients between marine and terrestrial systems by catching fish and marine invertebrates at sea and excreting guano on land. This distinctive mechanism of bird-mediated nutrient cycling is a key factor in maintaining productivity in many island ecosystems, thus enabling islands to support a diverse range of terrestrial species. As illustrated by Gough Island's albatrosses and Scotland's extinct gadfly petrel, the arrival of hungry rats and mice typically causes seabird colonies to collapse. This collapse not only disrupts terrestrial island ecosystems but also causes coral reefs to become depleted, as guano also runs off the land and supports increased productivity in nearshore ecosystems. Recent research in the Chagos Archipelago has found that, compared to islands with rats on them, nearby rat-free islands have 760 times the density of seabirds and over 250 times the amount of nitrogen in soil and plant samples. Reefs around rat-free islands also have faster-growing damselfish, and fish communities with almost 50 per cent greater overall biomass.[22] The arrival of invasive rodents on islands can therefore be catastrophic and far-reaching; species extinctions are just the beginning.

Even native island rodents are at risk from invasive rodents. This can be through competition, with aggressive generalist invaders consuming the same food resources and filling the same dietary niche as natives. The large Caribbean island of Hispaniola was formerly home to a diverse endemic rodent fauna including native spiny rats and numerous species of hutias. However, Hispaniola was the first place in the New World to be settled by Europeans – its name is a corruption of 'La Isla Española' or 'the Spanish island', as it was called by Columbus in 1492. Black rats probably arrived with Columbus and the European settlers who followed shortly afterwards, and centuries-old rat bones in caves show they spread rapidly across the island. My colleagues Siobhán Cooke and Brooke Crowley have reconstructed the diets of Hispaniola's native rodents and the first black rat invaders by studying the ratios of different isotopes of two elements, carbon and oxygen, in old rodent teeth from cave deposits similar to Liang Lawuala.[23] These isotopic signatures are influenced by the plants the rodents consumed when they were alive. Siobhán and Brooke revealed the invasive rats ate a broad range of plants that included the main foods of many native Hispaniolan rodents. Whereas black rats are still common across Hispaniola, the island's native inhabitants were not as lucky; like the Little Swan Island hutia once found further south in the Caribbean, all but one of Hispaniola's twelve spiny rats and hutias are now extinct and black rats are a prime suspect in these disappearances.

Competition is just one mechanism by which invasive rodents have eliminated native island rodents, as revealed by an elegant study involving ancient DNA research.[24] Christmas Island, an isolated landmass in the eastern Indian Ocean, was formerly home to two large native rodents, Maclear's rat and the wonderfully-named bulldog rat. When Christmas Island was colonised in the 1890s to mine the island's commercially important phosphate deposits (which had formed through accumulation of guano from seabird colonies), the native rats were said to be abundant. Of course, it wasn't long before black rats got to Christmas Island; it's thought they arrived aboard the SS *Hindustan* in 1899. Within a few years the familiar story played out and the native rats vanished, with neither species seen after 1905. It was initially thought that black rats might have diluted the natives away through hybridisation, as the three species were closely related. Interestingly though, the last reported sightings of native rats were of sick-looking animals crawling feebly along footpaths.

Fortuitously, a handful of specimens of both native rats and recently-arrived black rats were collected on Christmas Island by contemporary zoologists and preserved in museum collections in London, Oxford and Cambridge. These specimens include native rats collected both before and after the arrival of black rats. Ancient DNA analysis of fragments of dried skin taken from these archival specimens has provided an explanation for their disappearance. The study found no evidence of hybridisation between native rats and invaders, but did find something else – genetic evidence that the invaders were infected with a single-celled parasite called a trypanosome, which is carried by black rat fleas, and the same type of parasite that causes sleeping sickness and other fatal diseases in humans. No native rats collected before the arrival of the SS *Hindustan* showed evidence of infection, but native rats collected after 1899 contained the characteristic genetic signal of trypanosomes. Instead of outcompeting or hybridising the native rats, the invaders had brought a virulent disease against which their naïve native relatives had no resistance. Black rats and their fleas helped to spread the Black Death, the most fatal human pandemic in history, and it turns out they can act as disease vectors to endanger other mammals as well. Much like the human inhabitants of St Kilda and the Taino people of the Caribbean, the rats of Christmas Island were doomed by their epidemiological isolation.

Invasive rats proved a source of consternation in Mahaniwa, too. One night I'd been in the middle of a strange dream about getting a haircut when I suddenly woke up. It was still completely dark as I opened my eyes and wondered groggily what time it was. I must still be half-asleep; I thought I could still feel the barber's fingers running through my hair and pressing against my scalp. Suddenly I was completely awake and beating my head with my hands – there were rats crawling around on my pillow and tugging and gnawing at my hair. I pulled my sleeping bag over my head and held it tightly closed, listening to the yells from Tim and James as they rapidly became aware of what was happening too.

We carried on with our palaeontological research around Mahaniwa. No other caves yielded much of interest, but the rich fossil deposit in Liang Lawuala more than made up for this – the cavern was full of the remains of our new giant rodents, densely scattered over a relatively small area of cave floor. Further investigation revealed a probable explanation for this unusual concentration of bones. A couple of metres further down the tunnel the ceiling became a mess of broken slabs, through which material from the cave entrance – situated directly overhead – had fallen to form a cone of debris. The fossils were part of the surface detritus that had ended up in this subterranean tunnel, falling through cracks in the rock above as 'bone rain'. This sort of concentrated fossil deposit, typically found near a cave entrance, is characteristic of an ancient roost or nest site of an owl. The porch in the cliff face overlooking the Mahaniwa valley would have been an excellent owl roost and our giant rats would have been ideal prey. Owls have been hugely important agents of fossil deposition, and the preserved remains of ancient owl meals – in caves from the Caribbean to New Zealand – are often the only clue to help palaeontologists reconstruct prehistoric faunas. Indeed, they're honoured in one of my favourite titles for a scientific paper, written by Gerrit Miller in 1929: 'Mammals eaten by Indians, owls, and Spaniards in the coast region of the Dominican Republic'.[25] However, realising that the Liang Lawuala bone deposit was a prehistoric owl accumulation meant we could only learn about a subset of Sumba's past fauna from this site. Large rats are ideal food for a raptor, but other components of the fossil fauna found on nearby Flores – Komodo dragons, stegodons, and hobbits – are all much too large to become an owl's dinner. The bones of Sumba's ancient humans, be they pygmies or giants (and if indeed they had ever existed), could not have been preserved in this sort of fossil site – they would have to be searched for elsewhere.

Liang Lawuala did have more surprises for us, though. Sorting carefully through the bones we'd found, Jen spotted two distinctive jawbones that were definitely not from a rodent. These bones were unusually flattened, and instead of tiny mammalian molars they were lined with odd-looking rounded teeth that looked like they would crush rather than chew. We flicked through our battered photocopies of the old palaeontological literature about Flores and quickly found our suspect. These were jawbones of *Varanus hooijeri* – the weird extinct monitor lizard that might have eaten snails or hard fruit stones, and until now known only from a few caves on Flores, including the hobbit cave of Liang Bua! Whereas the giant rats from Mahaniwa seemed different to the rodents of Flores, these new fossils provided a direct link between the fossil faunas of Flores and Sumba. This was hugely exciting. What else was waiting to be found out here on this uncharted island?

Maybe a new species of owl, for starters. Four living owls are known from Sumba – two widespread species of barn owl and two endemic boobook owls, the imaginatively-named little Sumba boobook and great Sumba boobook (so-called because of their calls, although according to my bird guide the call of the little Sumba boobook was apparently just a 'single "poop"'[26]). The smaller boobook was only described as recently as 2002, highlighting how poorly known Sumba's modern fauna still was. However, even the two barn owls, the largest of the island's owls, have diets that typically consist of small rodents weighing under 100g – much smaller than the giant rats and monitor lizards we were unearthing. Fossil owl accumulations in the Caribbean that contain comparably large rodents (the now-largely extinct hutias) are known to have been deposited by now-extinct giant owls. Based on the relatively scant remains of the owls themselves, some appear to have been truly monstrous beasts – in the memorable words of my friend Jo Cooper, a curator of the Natural History Museum's bird collections, these were 'owls the size of dustbins'. So if the Liang Lawuala fossil deposit was made by an owl, was this accumulator actually another now-extinct giant island endemic, the bones of which were also waiting to be discovered somewhere in another cave on Sumba?

I got used to the daily routine of life in Mahaniwa. When we were not out exploring caves, we sat around outside the headman's house, or walked along the road beside the cliff and the creek and explored the landscape around the village. The track ended beside a beautiful megalithic tomb, the capstone of which was being used to dry coffee beans laid out on a cloth. The other direction led to a sheltered area where we could wash in the creek without being seen – although this sometimes backfired. One evening when I was communing privately with nature, a battered old bus appeared out of nowhere on the road above, full of people laughing and waving out of the windows and seemingly having a party onboard. The bus was just high enough that everyone onboard could see right down into the creek, and I became the unwitting entertainment for the highly amused and very attentive party as the bus trundled past. These random party buses would suddenly rattle along the dirt road past sleepy Mahaniwa every few days, music blaring. They were almost as big a mystery to us as the mili mongga, although they did seem a pretty fun way to travel across the island. We tried to time our washes in the creek to avoid any well-intentioned encouragement from the passengers.

I tried to practice speaking Indonesian. It was a fun language to learn, and I soon had a series of set phrases to trot out as required – *tikus besar dan gajah kecil* (giant rat and tiny elephant) and *saya tidak mau pergi ke gua-gua lagi* (I don't want to go to any more caves) rapidly becoming favourites. Luckily for me, Indonesian has to be reasonably easy to function effectively as a lingua franca or trading language across this vast country, and most Indonesians have themselves learned it as a second language – for instance, here in Mahaniwa everyone spoke Eastern Sumbanese as their native language. Indonesian is slightly artificial, containing numerous loanwords from Dutch, Portuguese and several other vocabularies, and was chosen as the national language because it wasn't really anyone's mother tongue and so wouldn't reinforce regional inequalities. Indonesian has fewer words than most languages, and it's grammatically very simple – for instance, to make a plural you just duplicate the noun, so *tulang* (bone) becomes *tulang-tulang* (bones), another favourite term of ours. After only a few days we were all rabbiting on at Umbu and the headman's family in clumsy broken snatches of Indonesian (except for Tim, who could actually speak the language properly and humoured our efforts like a slightly bored parent).

We spent a lot of time sitting outside chatting with Umbu. On the other side of the yard the ancient old woman sat folded all day over her coffee beans; the sleepy rhythmic *tok tok … tok tok* was the constant background to our conversations. One day during a break from the caves, on a whim I asked Umbu how old he thought we all were. His eyes narrowed in thought. He looked at Jen. 'Twenty-five,' he said. 'Twenty-five … twenty-five,' he continued, moving on to James and Tim. He turned to me, and considered. '…Sixty?' The eruption of uncontrolled laughter from the others – and the expression of hurt shock on my face – took him aback. He looked apologetic. 'Maybe… fifty-seven?' Jen was now crying with laughter. I was in my thirties. I mumbled something about the cultural impression of seniority associated with being Pak Sam the group leader and the only one with pale hair, which might be associated with old age in dark-haired Asian cultures, and tried to move the conversation along swiftly. I knew I wasn't going to be allowed to forget this. It was almost as bad as the time in the Dominican Republic several years earlier, when after a long day of fieldwork looking for rare mammals, a random man had followed me into the hotel lobby and decided to announce (in front of all my colleagues) that I looked 'like Sting, but fatter'.

The news about our discovery of giant rat bones soon spread across the village, and a couple of men arrived at the headman's house in excited conversation. Umbu explained that they knew of another cave, not that far away, where they went every now and then to hunt fruit bats. After they knocked a bat down from the cave roof with a stick, they said that sometimes a huge rat would suddenly rush out of the rocks and drag it away. Maybe this was the same beast that we'd found! The men offered to take one of us to explore the bat cave – which they said was a vertical cavity dropping down into the limestone. Suddenly the horizontal underground tunnel with the soiled sheet in Liang Lawuala seemed like a desirable option … and anyway, I was obviously far too old for this sort of adventurous gallivanting. Sprightly young Tim offered to go with Umbu and the villagers to check out the cave.

Several hours later, he returned carrying a grubby plastic bag that absolutely stank. He told us they had climbed down a bamboo ladder into the huge hole and crept into a long underground side-tunnel – and sat in the darkness for hours, waiting. Umbu sprinkled a few dried sardines onto the tunnel floor, and eventually a rat had emerged and started to feed. The photos that Tim had taken showed a completely typical black rat – the recent invader, not our newly discovered giant beast. Then Umbu had smelled something in the cave; there was a dead thing further down the tunnel. Together they had scooped the decomposing mess into a bag and brought it back for us to examine. We all agreed that Umbu and Tim had done more than enough for the day, so Jen and I took the bag to the edge of the village and poured out the contents – a slurry of fur, decaying guts and maggots. Under the hot tropical sun, the smell was indescribable. Pausing frequently to gag, we managed to locate a skull and jaws amongst the fetid furry soup. It was another bloody black rat. As we turned to walk back, a thin dog ran over and gobbled it up.

Luckily there was more local biodiversity to see than just black rats. Lorikeets and other colourful tropical birds darted back and forth over the forest canopy lining the cliff above the village. Huge purple and green metallic pigeons sat in the trees overhanging the cliff and boomed their soothing sonorous calls out across the valley. As evening fell, the birds were replaced by huge fruit bats, which cranked their leathery wings slowly through the air. Who knew what other animals might also survive in these remote understudied forests?

One day while climbing the slope to get to another cave – this one imaginatively christened 'Goat Cave', because it was caked with goat dung – we were disturbed by rustlings in the trees overhead. A series of tiny human-like faces peered out through the leaves, then vanished. We'd startled a troop of long-tailed macaques. Sumba is located in the Indonesian island chain far beyond Wallace's Line; monkeys are characteristic members of the tropical rainforest fauna of continental Asia and the adjacent Sunda continental-shelf islands, but are bad overwater dispersers. Within the biogeographically fascinating region of Wallacea, only Sulawesi is home to endemic monkeys – a remarkable evolutionary radiation of seven species of macaque is distributed across different parts of this huge oddly-shaped island, and might represent two separate overwater colonisation events from Borneo.[27]

The macaques on Sumba had a different origin. Monkeys have never naturally colonised the islands of East Nusa Tenggara – they were introduced by people. This seems to have happened a long time ago. Although Sumba's ancient history and prehistory is poorly understood, the recent fossil record of neighbouring Flores shows evidence of macaques accompanying the arrival of Neolithic settlers from the Sunda Shelf across Wallacea around 4,000 years ago, during Jared Diamond's 'Chinese steamroller' population expansion event. Macaques were accompanied by several other non-native mammals such as porcupines, pigs, and civets[28] – medium-sized carnivorous mammals that look like a cross between a cat and a raccoon, and famous in the West due to their penchant for eating coffee beans; the defecated beans supposedly make the world's best coffee (after they've been thoroughly cleaned). Some of these species were clearly transported between islands for food; pigs and porcupines have a lot of meat on them, and porcupines are regularly hunted on Flores today.[29] The reason for the prehistoric introduction of civets and monkeys is less obvious (it wasn't for making good coffee, since coffee is not native to Asia and plantations weren't established in Indonesia until the seventeenth century). They could have been eaten as well, although civets might also have been used to catch rats, whilst monkeys may just have been pets. Either way, feral populations of most of these mammals now run wild across Flores and Sumba, although for some reason porcupines were never successfully introduced to Sumba during prehistory. Indeed, underneath its surface layer of dung the sediments of Goat Cave yielded the ancient remains of a civet that might have been hundreds or thousands of years old. More recently other invasive mammals, notably rusa deer from Java, have also been introduced widely across the region.

Thanks to widespread historical and prehistoric introductions, long-tailed macaques now have the third-largest geographic distribution of any primate, second only to humans and rhesus macaques. And macaque introductions followed the now-familiar formula of what always happens when people move animals into new ecosystems – has this ever ended well for local biodiversity? Despite their alternative name of crab-eating macaque, suggesting a somewhat refined diet, these monkeys are voracious and eat everything – and cause ecological havoc when they arrive on primate-free islands. The International Union for Conservation of Nature lists them as one of the world's 100 worst invasive species, and they are at least partly responsible for the most famous historical extinction of them all. Long-tailed macaques were brought to the island of Mauritius in the early 1600s, probably as pets onboard Dutch ships crossing the Indian Ocean from Southeast Asia. Once there, macaques (along with invasive rats and pigs) rapidly consumed the eggs and nestlings of the island's birds, which had evolved without mammalian predators and were generally flightless and ground-nesting. This island avifauna included a huge flightless pigeon: the dodo. Although sailors and settlers hunted dodos too, the human population of Mauritius never exceeded 50 people during the seventeenth century; but macaques and other invasive mammals spread rapidly across the island and wreaked catastrophic damage. By the end of the 1600s, the dodo was gone.[30]

It's likely that long-tailed macaques have had similar negative impacts on island ecosystems across Wallacea – but we have such a limited understanding of historical baselines for these ecosystems, and how they've changed over time, that it's hard to tell. However, macaques might also have had unexpected influences on local stories and beliefs in these islands. People from many cultural backgrounds, including our own, have often been confused in the past by the physical similarities between humans, apes and monkeys, which we now know reflect relatively recent shared evolutionary relationships. In this socio-cultural context, primates represent liminal animals that sit at the boundary of what we consider human. Historical appreciation of the curiously familiar characteristics of apes and monkeys prompted speculation and debate about the nature, definition, and uniqueness of humanity, and non-human primates were frequently attributed with specifically human characteristics or even erroneously identified as people.

In his *Natural History*, Pliny the Elder described a series of fanciful tribes from around what was then the known world, including satyrs and the tiny Pygmaioi, who were engaged in an endless war with migrating flocks of cranes (and named from the ancient Greek word *pygme*, the distance from elbow to knuckles, referring to their diminutive size). Some of Pliny's tribes have little relationship with any reality, and some might be distorted accounts of real contemporary ethnic groups, but historical commentators such as the seventeenth-century Dutch anatomist Nicolaes Tulp suggested that others might have their origin in ancient reports of African or Asian primates. The first scientific dissection of a chimpanzee (referred to as a 'pygmie'), conducted by comparative anatomist Edward Tyson in 1699, was actually carried out specifically to demonstrate that 'the pygmies, the cynocephali, the satyrs, and sphinges of the ancients' were 'all either apes or monkeys, and not men, as formerly pretended'.[31] Pliny based his *Natural History* on older accounts such as that of the Carthaginian explorer Hanno, who reported travelling to an island off Africa inhabited by what were described as savage, hairy people called Gorillai. It's impossible now to interpret what these beings might really have been, but when gorillas were described scientifically in the nineteenth century they were named after Hanno's tribe. Indeed, some ethnic groups across Africa and Asia today – including the Indigenous peoples of the Kagwene Mountain region of Cameroon, the Bondango people of the Democratic Republic of Congo, or the Lisu people of south-western China – will not hunt great apes or gibbons because they regard these primates as ancestors, or categorise them as human rather than animal.[32] Confusion over the distinction between humans and other primates continued into later centuries in Europe. For instance, a live chimp known as 'Madame Chimpanzee', exhibited in London in 1738, was described by observers as 'perfectly of a human Specie'.[33]

Early European encounters in Southeast Asia provided further challenges to the distinction between humans and other primates, and what this might mean for understanding ourselves and our natural state – metaphysical concerns that vexed the great Enlightenment thinkers. Rousseau believed that 'orang-outangs' (a term then used for both Asian and African apes) were human, although he remained suspicious of travellers' accounts. He was correct to have doubts, as many philosophical conclusions about the human condition were based upon erroneous observations by overly credulous travellers. The Scottish scholar and proto-evolutionary theorist James Burnett, Lord Monboddo, proposed – almost a century before Darwin – that apes represented the earliest form of humanity, and we had somehow evolved from non-human primates;[34] differences between 'us' and 'them' were those of degree rather than fundamental category, and it was 'impossible to draw a line betwixt the Orang Outang and the dumb persons among us'.[35] However, this theory was based partly upon a report by Swedish sailor Nils Matsson Kiöping, who saw on one of the Nicobar Islands 'a race of men ... who had tails like those of cats, and which they moved in the same manner'.[36] It's now thought this might have been a mistaken observation of the widely-distributed long-tailed macaque, which occurs naturally on these islands.

Other accounts attributed human-like characteristics to Asian apes. The Dutch physician Jacobus Bontius, working in Java in the seventeenth century, wrote wryly that 'the Javanese claimed that the *Ourang-Outangs* could talk, but that they did not want to because they did not want to be forced to work',[37] although he personally considered this 'ridiculous'.[38] In 1727, Scottish sea captain Alexander Hamilton claimed he had seen an orangutan keep a fire burning, boil fish to eat with rice, and blow its nose and 'throw away the Snot with his Fingers'.[39] Another European traveller, Daniel Beeckman, recounted in 1718 that his informants on Borneo told him that orangutans were men who had been 'Metamorphosed into Beasts for their Blasphemy'.[40] Even Linnaeus believed Kiöping's account that men with tails inhabited the East Indies, and accepted Bontius' report at face value, stating that 'troglodytes' (his taxonomic mish-mash of African and Asian apes) had the ability to speak:

> *They have their own language, spoken as a hissing sound produced in the throat. Their language is extremely difficult for other people to learn, unless a long time is spent among them. In 'our language' troglodytes are known to be able to say only 'yes' and 'no'.*[41]

Most infamously, the people of Hartlepool in northern England are supposed to have hanged a monkey washed up after a French shipwreck during the Napoleonic Wars in the early nineteenth century. As the locals had never seen either a monkey or a Frenchman before, so the story goes, they thought this ship's mascot was a French spy and so sentenced it to death. There is no actual evidence to back up the tale; however, it is now a proud part of Hartlepool's history and identity, and someone dressed as the town football club's mascot H'Angus the Monkey has been elected as mayor three times.[42]

Given this historical confusion over the distinction between people and primates, it's sometimes suggested that recent reports of wildmen or 'mystery man-beasts' might also be misidentifications of known species of apes or monkeys. William Osman Hill, the Zoological Society of London professor who proposed that the Sumatran orang pendek might be a dwarf island *Homo erectus*, also dismissed many supposed orang pendek accounts as sightings of gibbons and other primates.[43] Karel Dammerman, the Dutch biologist who studied Sumba's mammals, was also intrigued by orang pendek reports but was not credulous; he identified supposed orang pendek footprints as tracks of sun bears and suggested the wildman might represent a folk memory of locally extinct populations of orangutans, which had formerly occurred across Sumatra but were already in decline a century ago when Dammerman was collating reports.[44]

Other alleged wildmen have turned out to have a similarly mundane explanation. An animal exhibited as a live yeti in a zoo in Shigatse, Tibet, in the 1950s was found to be a gibbon.[45] Mysterious-looking creatures mentioned by Gregory Forth as witnessed by another anthropology professor in the 1980s on a jungle trail on Sumbawa, the neighbouring island to Sumba and Flores, sound very like long-tailed macaques,[46] and the legendary 'Bukit Timah Monkey Man' of Singapore might also be based upon mistaken sightings of this species. And a 'mysterious' animal supposedly shot by a Swiss oil prospector in Venezuela in 1917 – suggested by anthropologist Georges Montandon to represent an unknown ape or even a prehistoric hominid survivor, and which he christened *Ameranthropoides loysi* – can clearly be seen in the one existing photo to be just a dead spider monkey propped up on a box, its tail either hidden from view or amputated. It's suggested that Montandon actively perpetrated the hoax because of his outdated view of human evolution, whereby local human races were thought to have each descended from different types of ape. As the New World is inconveniently ape-free, Montandon seemingly decided to fabricate evidence of a suitable local 'ancestor' for South America's Amerindians to support this misguided evolutionary framework.[47]

Dammerman was even sent a specimen of what was supposedly an infant wildman but turned out to be the doctored remains of a leaf monkey.[48] Similar fake wildmen made of monkey body parts have also been used to try to deceive naïve observers in the Himalayas and elsewhere.[49] Dammerman's fake may have been produced by a similar process to that described from Sumatra during the thirteenth century by Marco Polo:

I would have you know that those who profess to have brought pygmy men from the Indies are involved in great falsehood and deception. For I assure you that these so-called pygmies are manufactured in this island; and I will tell you how. The truth is that there is a sort of monkey here which is very tiny and has a face very like a man's. So men take some of these monkeys, and remove all their hair, with a kind of ointment. Then they attach some long hairs to the chin in place of a beard, threading them through holes in the skin so that when the skin shrivels the holes shrink and the hairs seem to have grown there naturally. The feet and hands and other limbs which are not in conformity with the human figure are stretched and strained and remoulded by hand to the likeness of a man. Then the bodies are dried and treated with camphor and other drugs, so that they appear to be human. This is all a piece of trickery, as you have heard. For nowhere in all the Indies or in wilder regions still was there ever seen any man so tiny as these seem to be.[50]

Possible connections have also been made between long-tailed macaques and Nage stories of the ebu gogo on Flores. Some of Gregory Forth's Nage informants described the ebu gogo as a kind of ape or monkey, and some of the ways that monkeys are described in other Nage folktales, notably their mumbling way of talking, are similar to descriptions of the ebu gogo. The destruction of the ebu gogo as described in Nage mythology also closely matches fables from elsewhere in Indonesia about how a wily farmer tricked monkeys that were destroying his crops. The farmer persuaded the monkeys to climb into a rambutan tree, after stacking dry wood around it on the pretext of helping them climb up – then set fire to the wood and burned the monkeys to death. All, that is, except for a pregnant female that escaped, again paralleling the twist at the end of the ebu gogo story. Although Forth notes that monkeys are otherwise portrayed differently to the ebu gogo in other folklore on Flores, it's possible that the ebu gogo myth was influenced by cultural exchange of folktales between islands, and might contain elements of stories about monkeys from elsewhere in Southeast Asia.[51] And Forth also reported that a putative ebu gogo relic he was able to find on Flores turned out to be part of a skull from a macaque.[52,*]

There were no orangutans this side of Wallace's Line, or any other large mammals such as bears or gibbons that could be mistaken for wildmen on Sumba. But could the local presence of long-tailed macaques – which have been misinterpreted elsewhere in Southeast Asia as some sort of weird 'not-quite-people' – also help explain the stories we'd heard? Maybe. However, although a few aspects of these stories were vaguely reminiscent of monkeys – notably the long tail reported for the meu rumba – overall this Sumbanese wildman was much less monkey-like than the ebu gogo. Most obviously, it was supposedly a giant, unlike the diminutive monkey-sized Floresian entity. It felt like there was more to the mili mongga than just monkeys, although I was still uncertain what that might be.

---

* More recently, Forth has reported other 'ape-man' relics from elsewhere on Flores that turned out to be skulls, bones and teeth of dogs, ungulates, and the giant rat *Papagomys*.

And the longer we spent on Sumba, we uncovered not only new fossils but also new layers to the mili mongga story. Chatting with Umbu and a couple of villagers one afternoon, I realised I should ask whether anyone in Mahaniwa knew anything about our new favourite wildman. The response was a surprise. 'There is a mili mongga grave in the village,' Umbu translated for me, after discussion with the others. 'Some local people looking for treasure excavated it in the 1960s or 1970s and found bones as long as a person's arm.' I looked to the headman, who made the same distinctive gesture we'd seen before, moving his right hand from shoulder to wrist of his left arm and back again to demonstrate how large the bones had been. They sounded much too big to be just macaque bones, like Forth's ebu gogo relic.

Although we'd heard about giant bones before in Miurumba, that account had been vague and part of a wider narrative explaining how the name *anak mili mongga* had come about; but here in Mahaniwa there were people who could remember having dug up a supposed mili mongga grave and seen such bones themselves. 'What happened to these bones?' I asked, trying not to let my excitement become too obvious. 'They were reburied,' explained Umbu. Fair enough, I reflected; it would have been too much of a stroke of luck for them to still be lying around somewhere for visitors to just take a look. But this new information was *so* tantalising. 'Could the grave ever be dug up again, to re-examine the bones?' I asked nonchalantly. Further discussion in Sumbanese ensued; this didn't seem like an option, so I dropped the subject and thanked the villagers. But – fossils… graves… bones; it was all very strange. It felt like these different discoveries were all beginning to converge in unexpected ways. I wondered what else we might find.

CHAPTER SIX

TULANG JUNKIE

MONSTERS. No longer extant.

Gustave Flaubert, *Dictionary of Accepted Ideas*

Back in Waingapu, we regrouped for a few days to sort through our new fossils and plan where to look next. We had all our meals at a little restaurant round the corner from the hotel, the only place nearby where it was possible to buy food. They had a deceptively long menu, but if we tried to order anything other than nasi goreng (fried rice mixed with vegetables, egg and chillies), the waitress would smile at us sadly and shake her head. '*Kosong*' – not available. It became a challenge to see if *anything* on the huge menu was available other than nasi goreng, but the gentle reply was always the same: '*Kosong*'. It really didn't matter though – the nasi goreng always hit the spot. I would wash it down with a brightly-coloured and over-sweetened synthetic drink that quenched my thirst perfectly in the baking tropical heat. In breaks from cleaning, studying and packaging the bones from Mahaniwa, we also popped out to the little market down the road, where I persuaded Jen to buy two new outfits for a couple of dollars each – a no-nonsense snazzy white pant-suit with 'Casual or feminine look' emblazoned down its length in large black letters; and a diaphanous dress bearing a massive printed photo of Taylor Swift, surrounded by technological terms spelled out in blue lettering: 'Computer data', 'Differential', 'Address book'. For some reason, neither outfit was worn regularly during fieldwork.

We made a series of day-trips to investigate leads about other possible sites within a few hours' drive from Waingapu. I already hugely liked and respected Umbu both personally and professionally, but we rapidly found that he was clearly the best person to look for fossils with on Sumba – he was somehow connected to pretty much everyone across the entire island. For any random village in the middle of nowhere that we wanted to visit, he would frown in thought for a moment and then tell us he had a friend who knew someone there, or a cousin who lived just down the road. I realised just how lucky we were to work with him.

Despite Umbu's networking skills and connections, however, it began to feel like we'd run out of luck. None of the new caves we found and explored – after scrabbling through thickets and along desolate dirt tracks under the baking sun – yielded anything interesting at all. And there were lots of caves. There was the giant hole in the ground, which everyone other than Tim was too scared to climb into using a home-made bamboo ladder (Jen and I peered gingerly over the edge and called out vague encouragements). There was the cave that some village kids told us contained a human skull, which turned out to be a rotten coconut under some bat dung. There was the cave that was 'sacred', which seemed to mean that no one could remember exactly where it was – and which we spent several hours looking for while trudging along the edge of a cliff above a genuine crocodile-infested swamp (which I thought only existed in *Boy's Own* adventures). And there was the cave we had to crawl through dry leaf litter to get to, where James nearly put his hand on a green pit viper – I've never seen someone shimmy backwards on all fours so quickly. All the while we were followed by gangs of friendly children from nearby villages, daring themselves to yell out 'Hello mister!' and then grinning behind their hands and running to hide amongst each other. Although it was July, usually at least one of them would be wearing a bright red pointed Christmas hat with white trimming.

In each village we visited, Umbu also asked whether anyone could tell him about the mili mongga; he was now firmly on its trail and not prepared to give up. In contrast to our luck with caves, pretty much everybody knew something about the local *raksasa* or giant. We began to piece a few more things together. Unlike the orang pendek, people never found mili mongga footprints, but they knew that mili monggas were around because they would find snail shells that looked like they'd been eaten. And, it turned out, Mahaniwa wasn't special – lots of villages had mili mongga graves. Generally they were purported to exist in the most remote villages we visited, a long way from the nearest road. Whatever the mili mongga was, its bones certainly hadn't been in short supply – not exactly what one might expect for either a bigfoot or a late-surviving *Homo floresiensis*. I really wanted to know what all this was actually about.

We drove further east towards Rindi, stopping on the way at another village where Umbu had heard about someone else who might know about bones in caves. We found his contact, a man called Agus, and sat down to chat on the raised bamboo platform in front of his thatched hut. Umbu chuckled as Tim went straight for the betel nut that was offered round, and translated the conversation for us. 'He says there is a cave near the village; two hunters found it when they were cutting wood in the forest about twenty years ago. There were bones lying around on the surface. He says it was a dead mili mongga.'

We all looked up. We'd come here because we were trying to find new fossil sites, but it seemed we just couldn't get away from hearing about the local wildman. The stories about mili monggas and meu rumbas had all just been fun and cryptozoological, an entertaining side-line to why we were actually supposed to be visiting Sumba; but might the mili mongga have more in common with fossils than I'd initially thought? 'The hunters found the body of the mili mongga and told Agus about it–' (there was a pause while Umbu clarified an apparently important detail) '–One of the hunters couldn't speak, but the other one told Agus. He visited two years later and saw the mili mongga bones.'

I asked how they had known it was a mili mongga. 'No one around here has ever seen a living mili mongga, but they all knew what it must be because it had big bones. He says the skull was two feet long, and its teeth were as long as a finger. It had four big fangs. The whole skeleton was there, and it was … six metres long.' I looked at Agus; he was clearly getting into the story. I got the feeling that there might now be some exaggeration going on. 'All the bones looked like human bones, but… *bigger*.'

'Is it possible to visit the cave?' I asked. Further discussion ensued. 'We would have to go to another village first, that's two hours from here. We would have to leave there in the early morning and go through the jungle on horses. We could get to the cave by afternoon or evening.' Jen and I were starting to discuss whether this trek was worth considering when Umbu added some further pertinent information. 'There are no bones in the cave anymore. His friend visited the cave recently and said they have disappeared. But he says there's an old man in another village, only thirty minutes from here, who also found some bones that were very strange.' We thanked Agus. Tim spat out the chewed remains of his betel nut and we headed back to the car to find the old man.

As we drove eastward to the next village, we talked about this latest tale. First the mili mongga stories had been about wildmen visiting villages long ago, then they had been about big bones in graves, and now they were about big bones in caves – and it was big bones in caves that we were here to look for, too. Some of the story had felt a bit embellished, but the parallels were striking; and the 'big bones' could have been very old rather than from a recently dead mili mongga (whatever *that* was). By chance, had we come across some local folklore that was actually an explanation for the origin of unusual fossils?

We now know that fossils are the preserved remains or traces of life forms that lived long ago, nearly all of which are extinct today. But this viewpoint is surprisingly recent. Extinction only became scientifically accepted as a biological process at the end of the eighteenth century, when the French naturalist Baron Georges Cuvier, an expert in comparative anatomy, demonstrated in 1796 that mammoth fossils were distinct from living African and Asian elephants and thus represented an extinct species.[1] Before this, European thinkers held that the possibility of extinction was incompatible with the dominant paradigm of the Great Chain of Being. This concept was first proposed by the Neoplatonists and became incorporated into biblical dogma; it was hugely influential in medieval theology and into the Renaissance and Enlightenment. This worldview perceived nature as a fixed and unchanging hierarchy, with rocks at the bottom, followed by subdivided layers of plants, animals and humans, above which were angels – and finally with God at the top. The suggestion that God in His plenitude would permit any of His creations to be destroyed, thus upsetting the balance of nature through their removal from the Great Chain of Being, was anathema to Christian belief.[2] Literal interpretation of Scripture also prompted the orthodox opinion that the Earth was only a few thousand years old (with Creation famously calculated by Archbishop James Ussher in 1650 as having occurred at nightfall on 22 October, 4004 BC), meaning there was very little time for extinction or other long-term biological change to have occurred. Ironically, eventual acceptance of extinction at the end of the eighteenth century was initially used to support a continued biblical worldview and deny the possibility of evolution, another idea that challenged the prevailing paradigm of an unchanging natural hierarchy of forms; the idea that organisms could die out seemed antithetical to the idea that they might instead change over time in order to survive. If the fossil record provided evidence of ancient species that were now extinct, then Cuvier considered that the history of life must have consisted of a series of divine creations, each followed by a cataclysmic extinction where life perished, eventually leading to the events described in Genesis.[3] An example of how such beliefs influenced contemporary thinking is given by Thomas Jefferson, who wrote in the 1780s that 'Such is the oeconomy of nature, that no instance can be produced of her having permitted any one race of her animals to become extinct; of her having formed any link in her great work so weak as to be broken.'[4]

Before Cuvier demonstrated the possibility that species could become extinct, earlier thinkers had proposed a bewildering range of explanations for the fossilised remains of extinct life forms. 'True' fossils were typically grouped together with other rocks and minerals under the broader category of 'anything dug up from the ground' (the literal meaning of 'fossil', from the Latin *fossilis*). Indeed, fossils provide a salutary lesson about how paradigms change when confronted with new evidence – or how they don't. Consideration of fossils did eventually lead to a new understanding of the past, and some intellectual stumbling blocks were understandable, such as confusion over the process by which bone or shell could become replaced by stone. However, these enigmatic traces of prehistoric worlds were not recognised instinctively for what they were, but were instead incorporated (with surprising ease) into existing views about nature.

Many classical thinkers considered that fossils might represent the remains of long-dead animals and plants, with marine fossils found on land because the sea must periodically inundate the continents. However, although some later scholars such as Leonardo da Vinci continued to advocate an organic origin of fossils based on their own empirical observations, viewpoints had changed by the time that fossils vexed medieval, Renaissance and Enlightenment thinkers. These 'formed stones' were instead widely held to be the product of forces within the Earth trying to reproduce life inside rock. Aristotelian philosophy regarded all matter as alive, and that even stones could grow. Fossils could thus surely be produced within the Earth via spontaneous generation, from 'lapidifying juices' or 'wet exhalations'; or from more mysterious astral emanations, some sort of plastic affinity, or an 'aura seminalis'. This was similar to how salt, coal and ores were thought to grow in the ground, and how corals (then considered fully inorganic) were supposedly minerals that grew in the sea. Neoplatonic and Hermetic philosophy thought fossil formation was influenced by occult forces that shaped everything on Earth, with fossils the sympathetic reflections of the world soul and Plato's eternal forms.[5] Edward Lhuyd, an early keeper of Oxford's Ashmolean Museum, proposed the imaginative theory that fossils grew from the spawn and semen of marine animals that was exhaled from the ocean and fell to earth with rain and fog, and these impregnated fluids grew underground with the substance of rocks but the form of animals.[6] Or maybe fossils were simply sports of nature? Philosophers such as the Jesuit polymath Athanasius Kircher thought 'the great architect, as if in jest, had imitated the teeth and bones of animals, shells or snakes'.[7] Some thinkers simply ignored them as inconvenient – Thomas Jefferson, again not looking good in hindsight, considered many fossils to be 'beyond the investigation of human sagacity'.[8]

And fossils were often slotted conveniently into a biblical worldview, primarily as empirical evidence of the Deluge. Seashells found in unlikely places such as mountains were often explained away in this manner (although Voltaire proposed that they might instead be the discarded meals of pilgrims and crusaders returning from the Holy Land), and mammoth bones and ivory must be the remains of elephants washed northward from tropical regions by the force of the biblical Flood. John Woodward, an early Fellow of the Royal Society, suggested in 1695 that marine shells were distributed widely across terrestrial sediments because:

> ... *the whole Terrestrial Globe was taken all to pieces and dissolved at the Deluge, the Particles of Stone, Marble, and all other solid Fossils dissevered, taken up into the Water, and there sustained together with Sea-shells and other Animal and Vegetable Bodies; and that the present Earth consists, and was formed out of that promiscuous Mass of Sand, Earth, Shells, and the rest, falling down again, and subsiding from the Water.*[9]

At least he admitted his theory 'will perhaps at first sight seem very strange, and almost shock an ordinary Reader.'[10] The metre-long fossil of a giant salamander found in 1726 was thought by Swiss scholar Johann Scheuchzer to be the skeleton of a godless sinner who drowned in the Flood, which he described as *Homo diluvii testis* ('Man, witness of the Deluge'). Scheuchzer also wrote a pamphlet entitled 'Complaints and Justifications of the Fishes', an account of fossil fishes (led by a large pike from Lake Constance) complaining they had died innocently in the Deluge and were upset that laypeople mistook them for freaks of nature.[11]

Even those who correctly associated fossils with the remains of dead organisms often underestimated their antiquity, and had a limited frame of reference for speculating about what sort of animals these bones might have come from. Before Cuvier differentiated mammoth bones from those of living elephants based on their anatomical characteristics, mammoth bones found across Europe and Siberia were frequently thought to be the remains of Hannibal's Carthaginian war elephants – or the lost war elephants of Alexander the Great, according to Russian emperor Peter the Great.[12] The seventeenth-century linguist Georg Stiernhielm considered that the Finns, Lapps and Estonians, with their unusual non-Scandinavian languages, were actually the ten lost tribes of Israel, and mammoth remains were elephants that the lost tribes had apparently used to travel to the far north.[13] A huge thigh-bone found in Oxfordshire was also concluded by Robert Plot in 1676 to 'have belong'd to some greater *Animal* than either an *Ox* or *Horse*; and if so ... in probability it must have been the *bone* of some *Elephant* brought hither during the Government of the

Romans in *Britan'*.[14] This thigh-bone was later recognised to belong to a *Megalosaurus*, the first dinosaur to be scientifically described (although before that, in 1763 it was also famously given the name *Scrotum humanum* due to the broken specimen's similarity in general shape to male genitalia). Mammoth bones were interpreted as out-of-place elephants by the great eighteenth-century French naturalist Georges-Louis Leclerc, the Comte de Buffon, but he considered them evidence for his 'cooling Earth theory' – proposing the world had once been warm enough to support tropical animals at high latitudes, and these animals had shifted their distribution and 'degenerated' as temperatures dropped. Fossilised bones of a mastodon (an extinct mammoth-like animal) found in Ohio, described as the 'American Incognitum', were also considered by Buffon as merely a locally-vanished northern form of elephant.[15]

Others instead thought these unusual fossilised creatures must still survive somewhere in a remote corner of the Earth – a potentially plausible idea when much of the world remained unexplored. Thomas Jefferson, that long-time denier of extinction, almost certainly viewed the famous mounted skeleton of the American Incognitum in Charles Willson Peale's museum in Philadelphia, the first natural history museum in the United States (where it was displayed alongside a 36kg turnip and the trigger-finger of a convicted murderer), and instructed Lewis and Clark to search for living mastodons during their famous expedition across the American West. Referring to the mastodon, he wrote that 'He may as well exist there now, as he did formerly, where we find his bones.'[16] Other scholars who denied the possibility of extinction, such as Thomas Molyneux and John Ray, erroneously thought that bones of the extinct giant deer or 'Irish elk' *Megaloceros* were the same as the living American moose or European reindeer, or that ammonites surely survived in the depths of the ocean – these forms *must* all still persist, as literal 'living fossils'.

Whereas the connection between fossils and prehistoric life was only understood relatively recently by western thinkers, however, fossils themselves have been encountered and appreciated throughout human history. The Iron Age inhabitants of southern England collected fossil oysters, ammonites, and the vertebrae of extinct marine reptiles called ichthyosaurs, perforating them for possible use as spindle whorls.[17] The Neolithic long barrow at Stoney Littleton in Somerset, constructed 5,500 years ago, has a beautiful large ammonite fossil preserved in one of the stone lintels at its entrance, which was clearly chosen for this distinctive feature. Even further back in time, 15,000 years ago Palaeolithic people living in caves around what is now Arcy-sur-Cure in France hand-drilled a hole through the tail of a 400 million-year-old fossil trilobite, an ancient extinct woodlouse-like marine animal, presumably to wear it as an amulet. The cave in which this treasure was found is now known as the Grotte du Trilobite.[18]

We will never know what such fossils signified to our distant ancestors during these ancient encounters with deep time. However, historical accounts and anthropological observations show that fossils also inspired a range of interpretations from local commentators and Indigenous and rural peoples. Some local beliefs about fossils merely associated them with living animals or viewed them in utilitarian terms. Ammonites were snakes turned to stone by medieval saints such as St Hilda of Whitby or St Keyna of Keynsham; and the Greek philosopher Strabo reported that the fossilised remains of giant single-celled organisms called nummulites, visible in quarried stone around the Pyramids, were petrified lentils from the meals of ancient Egyptian workmen. Dinosaur footprints found in nineteenth-century New England were locally referred to as the marks of 'Noah's ravens', and other fossil dinosaur vertebrae and giant shellfish were just petrified horses' hooves.[19] Fossils were also widely prized for their supposedly medicinal properties, with different types used to treat numerous medieval ailments: 'glossopetrae' or tongue stones from Malta, actually fossilised shark teeth, were prized as an antidote to poison,[20] and so-called 'dragon bones' and 'dragon teeth' have been important in traditional Chinese medicine for thousands of years (remember how von Koenigswald discovered the teeth of *Gigantopithecus* in a Hong Kong drugstore), associated with the 'popular belief that dragons shed their bones as often as snakes shed their skins'.[21] The remains of ancient submerged Welsh forests, exposed at low tide as semi-fossilised tree stumps, were historically burned as poor-quality firewood,[22] and fragments of older petrified wood were utilised by prehistoric hunter-gatherers to make tools in the Sahara.[23] And sometimes large fossils were even used as building materials – petrified tree trunks were used to line roads in ancient Egypt,[24] and the explorer Roy Chapman Andrews found a shrine in the Gobi Desert constructed from the fossilised bones of *Indricotherium*, the largest-ever land mammal.[25]

The most poetic of such stories involves trilobites. These ancient extinct arthropods were thought to be 'bat stones', 'butterfly stones' or 'swallow stones' in China, and around Carmarthen in Wales they prompted a story about how Merlin met his end by falling in love with a fairy. In the words of the nineteenth-century geologist William Symonds:

> *One summer's day when the birds were singing, and the butterflies flitting, the wizard and the fairy entered a rocky cave, and here by the aid of a spell taught her by Merlin himself, the fairy closed the cavern and entombed the magician and the butterflies. Thus Merlin was "lost to life, and use, and name, and fame", and hence the appearance of the butterflies (or trilobites' tails) in the rocks of Mount Pleasant.*[26]

Other local interpretations of fossils imbued them with magical potency and protective magic, viewing them as charms or talismans or as having connections with the spirit world. Fossil sea urchins, which are shaped somewhat like little round loaves of bread, were called fairy loaves in eastern England and placed on mantelpieces in rural cottages to ensure there would always be bread in the house, or next to the oven to induce the bread to rise through sympathetic magic.[27] Like the Palaeolithic people of Arcy-sur-Cure, the Native American Pahvant Ute tribe of Utah wore trilobites round their necks as lucky charms, believing they could stop bullets, and aboriginal Australians have also carried charms made of the fossilised teeth of long-extinct giant marsupials.[28] Fossil oysters were called 'devil's toenails' in rural England, and the fossil skull of an ichthyosaur found in Kilve, Somerset was said to be a local dragon called Blue Ben, that had been the devil's steed until it was bogged in a mire and drowned.[29] I've also heard the disconcerting recent story that a stegodon fossil was donated to a small local museum in Southeast Asia, but was thought to have brought a ghost with it into the museum, so was hastily buried outside under some concrete.

From rural Ireland to New Guinea,[30] prehistoric hand axes and flint arrowheads were widely thought to be either petrified thunderbolts, or 'fairy darts' or 'elf shot' with magical powers and the ability to either cause or cure illness in people and cattle.[31] Shakespeare warns of such 'thunder-stones' in *Julius Caesar* and *Cymbeline*. In Java, stone axes were called *gigi gledek* – the teeth of lightning – because lightning was thought to be a huge animal that hides behind clouds and bites trees, losing its teeth in the process.[32] I myself have seen an ancient Amerindian hand-axe given pride of place as a supernaturally powerful ritual object on the voodoo altar of a local bokor or witch-doctor in the mountains of Haiti. Belemnites (the cone-like internal shells of extinct squid-like animals), which look superficially similar to sharp stone tools, were also widely thought to be elf shot. Their stiff, rod-like shape caused them to be regarded as gnome candles in Sweden; and, more suggestively, as a stand-in for the erect penis of the priapic fertility god Min in predynastic Egypt. In past centuries they were sold as 'Lynx Stone', a substance with magical curative properties but which could also make your skull explode, supposedly formed from the petrified urine of a lynx (which grows into a stone 'as soone as it is pist out', in the word of Dioscorides[33]). This unusual substance was eulogised due to the allegorically selfish behaviour of the lynx, which was said to bury its piss in the sand to hide it from people: 'They do this from a certain constitutional meanness, for fear that the piss should be useful as an ornament to the human race.'[34] The medicinal properties of lynx-urinated belemnites were even celebrated in medieval verse:

> *Voided by lynxes, to a precious stone*
> *Congealed the liquid is Lyncurium grown;*
> *This knows the lynx and strives with envious pride*
> *'Neath scraped-up sand the drops to hide …*
>
> *And let the patient wear the gem, its force*
> *Will soon arrest the diarrhoea's course.*[35]

Interestingly, neither the ancient Greeks nor many non-western peoples had any cultural barriers based on religious scripture to accepting the concept of extinction, and some of their stories about fossils are broadly consistent with the idea of species going extinct during prehistory. For instance, large fossil bones found on the Greek island of Samos during antiquity were interpreted as the remains of huge beasts called the Neades, which lived during primordial times and were swallowed by the earth for making too much noise.[36] The Yukaghir, an Indigenous people from the Russian Far East, thought that mammoth remains found in the permafrost were animals called *xolhut* that had existed long ago, but disappeared because they ate all the trees in northern Siberia and turned the landscape into tundra; 'the creation of the mammoth was a blunder of the Superior Being.'[37] Other cultures also believe that some animals declined and disappeared in the past, although this can form part of a cyclical worldview – the animals are not gone forever but will reappear again, as in the cosmology of the Cree people of North America.[38]

However, these legends also demonstrate how cultures that lacked a frame of reference for understanding deep time or evolutionary history have often interpreted large fossil bones in a specific way – they were the remains of monsters. It was standard for children's dinosaur books of the 1970s and 1980s to include a special feature page about these mistaken identities. One of the best-known examples is that of the Cyclopes, the monster race of one-eyed giants. The famous story of Polyphemus the Cyclops in Homer's *Odyssey* is from the Mediterranean, a region that contains many fossils of tiny dwarfed island elephants that existed a few tens of thousands of years ago. Elephant skulls have an extremely large merged nasal opening, so fossil skulls of extinct Mediterranean dwarf elephants could potentially have been mistaken for huge human skulls with a single central eye socket. Sir Hans Sloane, founder of the British Museum, suggested that a skeleton discovered in Sicily in 1342 and attributed to Polyphemus by Giovanni Boccaccio was probably that of an elephant.[39] This idea was promoted in the early twentieth century by Austrian palaeontologist Othenio Abel.[40,*] More recently, classical scholar Adrienne Mayor has popularised the idea that much of the ancient Greek pantheon of marvellous creatures was inspired by encounters with fossils, suggesting in particular that myths

---

* Unfortunately Abel also had some more unpleasant ideas, and regarded the day that the Nazi flag was flown over the university in Vienna as the happiest moment of his life.

of gold-guarding griffins might be based upon Scythian tales of central Asian *Protoceratops* dinosaur fossils that were found eroding from the ground near gold deposits.[41]

Fossils have also inspired myths and legends of monstrous beasts across many other parts of the world. Huge fossil bones found in South Dakota, Nebraska and Wyoming were thought by the Native American Oglala Sioux tribe to be the remains of 'thunder beasts' or 'thunder horses', which would jump from the clouds during thunderstorms and drive herds of bison towards the Sioux to be hunted. These bones were actually the remains of extinct rhinoceros-like mammals that were fittingly named brontotheres (literally meaning 'thunder beasts' in Greek) by palaeontologist Othniel Charles Marsh. Some Siberian tribes thought mammoths were giant mole-like creatures that burrowed underground with their tusks and were killed by fresh air (hence mammoth carcasses sometimes found sticking out of the ground).[42] Dragons in Chinese culture were sometimes depicted with the antlers of fossil deer species, and elephant fossils found in Japan were initially thought to be bones of dragons, although Japanese scholars correctly identified them relatively early on.[43] The skulls of woolly rhinos and cave bears found in European caves during the Middle Ages were also regularly mistaken for the remains of dragons in their lairs, becoming the basis for locally-celebrated monsters such as the 'lindwurm' of Klagenfurt in Austria. Even into the seventeenth century, Athanasius Kircher stated that he believed in dragons because he had seen their bones, cave bear skulls were described as those of dragons ('Draco') by the medical doctors János Paterson Hain and Heinrich Vollgnad, and a woolly rhino skeleton was famously reconstructed as a bizarre-looking unicorn in Magdeburg, Germany.[44]

And large fossil bones were often thought to be one thing in particular – the bones of giants. In addition to the monstrous Cyclopes, the bones of mammoths, extinct elephants and other large fossil mammals were frequently taken as evidence for the existence of a race of huge people that lived during antiquity, and were sometimes even identified as specific individual giants known from ancient texts or legends. Early commentators lacked knowledge about comparative anatomy and morphology (how the bones of different types of animals differ from each other), so were unable to appreciate the rather obvious differences between bones and teeth of humans and elephants. Classical authors such as Pliny and Herodotus referred to the discovery of huge bones of giants known from mythology such as Orion, Orestes and Ajax, and European scholars from past centuries added further finds of other famous giants such as Gog and Magog to this list. Fossil bones of a deinothere (a prehistoric elephant-like large mammal), found in southern France in 1613, were exhibited widely around Europe as the supposed remains of Teutobochus, the legendary giant king of the Teutons defeated by the Romans in 102 BC.[45] And giant bones were thought to be the bones of giants elsewhere in the world as well – for example in the Valley of Mexico, where mammoth fossils were interpreted by the Aztecs as proof that giants had once lived there in the distant past; and maybe also in India, where the rich mammalian fossil deposits of the Siwalik Hills might have inspired tales about battles of gigantic heroes in the legendary epic *Mahâbhârata*.[46]

From the sixteenth century onward, critics such as Johannes Goropius Becanus suggested that the bones of the supposed giant of Antwerp, or the bones and teeth of King Og found in a cave near Jerusalem, were actually just 'some old Elephant'.[47] This perceptive reassessment was sometimes backed up with the somewhat more speculative suggestion that buried bones and teeth can grow over time within the soil, in ground-water rich in 'effluvia'. The example of Ambroise Paré, an early pioneer in surgical techniques, was used as evidence – Paré allegedly put one of his own teeth into a crate in an underground room, and said it doubled in size after a few years.[48] However, even these reappraisals were not enough to convince other writers such as Reverend Agostino Calmet, who wrote in 1730 that 'It could be truth, that some bones of Elephant or whale are shown as bones of giants; but it is most certain, that in many places real bones of Giants are preserved, and consequently the existence of the giants is a fact beyond all doubt.'[49] New evidence and new theories were unable to overturn stubborn old paradigms.

It's uncertain whether discoveries of giant fossil bones inspired widespread beliefs about giants, or whether such discoveries just reinforced and perpetuated existing myths and legends. However, there was a surprisingly close connection between fossils – especially elephant fossils – and world folklore about giants and other monsters, which raises an intriguing new possibility about the origin of the mili mongga stories we were hearing. The one extinct animal known from Sumba before our trip was the tiny stegodon from Watumbaka, *Stegodon sumbaensis*: a fossil elephant with bones very similar to those regularly mistaken for bones of giants. So was the mili mongga actually what we'd come to Sumba to look for all along – was it just an old dead stegodon?

And fossils might not only be the explanation for supposed mili mongga bones in caves, but also for supposed mili mongga bones in graves. Throughout history people have misinterpreted large fossil bones as being those of monsters, and have also venerated them as relics. Bones of the sheep-sized extinct dwarf hippopotamus of Cyprus were long thought to be the remains of either St Phanourios or a group of pious early Christians known as the Seven Sleepers (although they were also regularly ground up for medicine).[50] When these hippo fossils were described scientifically, they were named *Phanourios* after this legend. William Buckland, the eccentric nineteenth-century palaeontologist who conducted the first cave palaeontology in Kirkdale Cavern, also encountered fossils masquerading as the bones of a saint during his honeymoon in Sicily in 1826. Buckland 'was less than impressed by some bones exhibited as the remains of St Rosalia'; he 'promptly identified them as belonging to a defunct goat.'[51] However, 'this fact caused not the slightest diminution in their miraculous power.'[52] The sacredness of fossils was not restricted to western cultures. Fossil shark teeth, thought to be talons of the long-nosed goblin *tengu-no-tsume*, were enshrined in temples in Japan.[53] Even the first known jawbone of a Denisovan – an enigmatic, recently-extinct species of human from central Asia, which was a contemporary of Neanderthals, hobbits and early modern humans – was found by a Buddhist monk in a sacred cave in northern China where local people collected 'holy bones' to use for medicine, and was originally given to the Sixth Gungthang Rinpoche or Living Buddha.[54]

Until only a few hundred years ago, churches across Europe regularly kept mammoth bones as evidence of the reality of biblical giants or behemoths alongside other dubious relics. Giant fossil bones thought to be the remains of saints were not only kept above-ground in reliquaries, but were also placed in coffins and reburied in medieval churches or tombs across Europe.[55] This practice also occurred throughout classical times and is documented from as early as the seventh century BC.[56] A few centuries later, the Roman Emperor Hadrian, while visiting the ruins of Troy, was shown some huge bones found eroding out on the coast that were thought to belong to the colossal Greek warrior Ajax; he was so alarmed by the giant skeleton that he ordered a tomb to be built so it could be properly reburied.[57] Even prehistoric human fossils were sometimes given a proper Christian burial in the local cemetery, such as the seventeen Cro-Magnon skeletons found interred with mammoth teeth near the French village of Aurignac in 1852.[58] Indeed, some historical commentators considered that placing large animal bones into graves alongside human corpses was an ancient ritual.[59]

These ancient fossil burials were a source of historical confusion. Robert Plot, pondering the identity of the mysterious scrotum-like dinosaur bone found in Oxfordshire, also observed that 'since the great conflagration of *London*, *Anno* 1666. upon the pulling down of St. *Mary Wool-Church* ... there was found a *thigh-bone* (supposed to be of a *Woman*) now to be seen at the *King's-head Tavern* at *Greenwich* in *Kent*, much bigger and longer than ours of *stone* could in proportion be, had it been intire'. In addition to a giant tooth he had also obtained, 'there were two others ... dug up in the Parish Church of *Morton Valence*'. Plot again thought these were the remains of elephants, and was puzzled by their connection with churches: 'Now how *Elephants* should come to be buryed in *Churches*, is a question not easily answered, except we will run to so groundless a shift, as to say, that possibly the *Elephants* might be there buryed before *Christianity* florish'd in *Britan*, and that these Churches were afterward casually built over them.'[60]

So could stories about mili monggas have been inspired by local discoveries of fossil elephants or other animals, just like the bones we were here to look for ourselves — and might these bones even have been reburied and venerated as the remains of giants? It was certainly a fascinating and unexpected possibility. In his book *Why the Porcupine is Not a Bird: Explorations in the Folk Zoology of an Eastern Indonesian*

People, Gregory Forth similarly proposed that subterranean buffalo-like creatures called *bhada sula* and *ana gezu*, from the mythology of the Nage and other ethnic groups on Flores, might be based upon local discoveries of stegodon bones (or maybe their tusks, which superficially resemble buffalo horns).[61] These beasts are not that dissimilar to the giant subterranean creatures imagined from mammoth remains by Siberian tribes. Forth raises a note of caution that he could not confirm whether such stories were definitely ancient, or instead potentially derived from knowledge of palaeontological discoveries from the twentieth century; however, he also observed that local people on Flores were generally uninterested in western scientific reporting of local fossils such as *Homo floresiensis*.[62] Interestingly, stories of the *ana gezu* also maintain that it was wiped out by human hunters. So if stegodon fossils might have inspired legends of now-extinct monsters on nearby islands in Wallacea, might they have inspired legends of giant wildmen on Sumba too? Did this mean we needed to chase mili monggas in order to discover fossils?

The old man who was supposed to know about strange bones was hard to find; we had to stop a couple of times for Umbu to yell out for directions from people working in their yards. Eventually we bounced down a dirt track and pulled up at a tiny dilapidated hut half-hidden in the scrub. A young boy was standing outside, absent-mindedly sucking on the sharp end of a pair of filthy-looking scissors as he watched us climb from the car. In his right hand was a live chicken, dazed and hanging upside-down by the legs. A skinny cat crouched next to him, staring intently at the bird and reaching up with its paws to slash at its head. We waited awkwardly, feeling the scrutiny of the boy's impassive gaze as he moved the scissors around in his mouth. Umbu called into the hut; there was a faint reply and he stepped inside, gesturing for me to follow.

It took a moment for my eyes to get used to the darkness inside. There were no belongings or furniture; on the floor, lying quietly under a thin grey sheet, was the very old man. It felt intrusive to be here, asking about stupid fossils in the face of this poverty. I wanted to leave but the man was trying to tell Umbu about the bones he'd seen. His voice was barely a whisper. Umbu looked down. 'He's forgotten the details. He says they were just the bones of a cow or buffalo. They were just lying around on the ground somewhere, not in a cave.' I thanked the old man sincerely and apologised for disturbing him as he laid his head back down onto the floor and closed his eyes.

We left the hut and walked quickly back to the car. Another man in dirty clothes was sitting on a low wall and leaning forward on a stick. He turned his face towards us and I realised he was blind. We drove quietly away. I was privileged and lucky; the people here had more pressing things to be dealing with than old bones or extinct species.

We reached Rindi as the sun was setting. It was a beautiful village, with majestic traditional thatched Sumbanese houses towering up on both sides of the grassy open area where we parked. At the end of the rows of tall houses was a group of impressive megalithic tombs. Umbu took us over to point out one in particular. It was rapidly getting dark, and the palm trees around the houses and graves were now just silhouettes against a blue-black sky. On top of the huge horizontal slab of one of the tombs was a curious carved stone figure, hard to see properly in the inky light. It depicted someone, or something, hunched forward on its knees, with arms pressed against its sides and a round flattened face staring blankly ahead. It was squat, with broad features and carved eyes that had been emphasised with charcoal or other black pigment. 'This is a mili mongga grave,' said Umbu.

There were three local stories to explain why the grave was here, Umbu told us. 'Maybe when the first people arrived in Rindi to build the village, there was a mili mongga here, so they killed it and buried it. Or maybe someone killed a mili mongga and built a tomb as a memorial – but it doesn't contain bones. Or maybe, a long time ago someone came to Rindi from another place where there were lots of mili monggas, and he built the grave as a memorial to them. Nobody knows.' Here was another twist. We'd heard about lots of mili mongga graves in the countryside across Sumba, but maybe these were actually memorials rather than graves; maybe only a few – or none at all – really contained any long-dead mili monggas, be they fossilised stegodon bones or something else entirely. The answer seemed as far away as ever.

It was now completely black outside; I could just make Umbu out as a shape in front of me. 'Would you like to meet my aunt?' he asked.

'She is about to have her dinner.' We followed him into one of the large thatched houses nearby. Several people were already inside sitting against the walls, and we were introduced to various members of Umbu's extended family with lots of smiles, handshakes and warm greetings. Umbu told everyone why we were here – by now he just said we were interested in mili monggas rather than fossils – and an uncle explained to us that he always took his dogs with him when he went into the forest, because mili monggas were scared of their barking.

In the middle of the room, on a large rattan mat stretched across the floor, was a sort of open-topped enclosure about a metre and a half across. Its sides were draped with beautiful ikat textiles that carried complex white and black geometric designs, and in the middle was an upright figure entirely covered by a shiny purple cloth. Umbu gestured respectfully towards the enclosure. 'There is my aunt,' he told us. Oh, I realised. She was dead.

Everyone in the room became silent, and a small woman moved out in front of the enclosure. Here, on a low table covered with more ikat fabrics, was laid out a series of dishes and cups. The woman reached down to a woven basket beside the table, and placed areca seeds, crushed lime and betel catkin together into a small bowl in her hand. She quietly ground the *sirih pinang* ingredients together with a small pestle, held them out in her hand towards the dead woman under the shawl, then poured them away between cracks in the wooden floor. Next she held out a small cup and poured that away too, followed by an offering of food. Umbu's aunt, still a part of the family, was being given dinner. I hadn't expected to be invited to share in such an intimate and private experience. I was humbled by this mark of friendship.

Umbu explained his aunt had been dead for seven years, but would not be interred until there was sufficient gold and rich ikat textiles to bury her with, according to the rules and customs that regulated how funerals were conducted. It was not uncommon for deceased family members to remain in the house for months or years, until their living relatives had accumulated enough money for an expensive tomb, grave goods, and a funeral feast and buffalo sacrifice. The customs were similar to the elaborate funeral rituals of the Torajan people from the highlands of Sulawesi, the larger island to the north. Gold coins or pendants were placed inside the mouth and hands of the dead person, and they were fed daily before sunrise and just after sunset. Cooking and weaving were forbidden in a house containing a dead relative, and the fire had to be kept burning. Here death was not a sudden event, but a gradual transition that was only concluded upon burial, and the living were expected to interact with dead family members, who were kept in the house until this final end was reached. Gregory Forth wrote that during his graduate research on Sumba, 'my wife and I were often invited to chew betel with someone awaiting burial.'[63]

It had been a long and unexpected day, with life and death close together everywhere we went. Each day here seemed to come with new surprises, both exciting and sobering. What new discoveries would we make about fossils – and mili monggas – tomorrow?

CHAPTER SEVEN

THE WALL OF THE MILI MONGGA

... the natives of Sumba, an east Indian island, celebrate a New Year's festival, which is at the same time a festival of the dead. The graves are in the middle of the village, and at a given moment all the people repair to them and raise a loud weeping and wailing. Then after indulging for a short time in the national pastimes they disperse back to their houses, and every family calls upon its dead to come back. The ghosts are believed to hear and accept the invitation. Accordingly betel and areca nuts are set out for them. Victims, too, are sacrificed in front of every house, and their hearts and livers are offered with rice to the dead. After a decent interval these portions are distributed amongst the living, who consume them and banquet gaily on flesh and rice, a rare event in their frugal lives. Then they play, dance, and sing to their heart's content, and the festival which began so lugubriously ends by being the merriest of the year. A little before daybreak the invisible guests take their departure. All the people turn out of their houses to escort them a little way. Holding in one hand the half of a coco-nut, which contains a small packet of provisions for the dead, and in the other hand a piece of smouldering wood, they march in procession, singing a drawling song to the accompaniment of a gong and waving the lighted brands in time to the music. So they move through the darkness till with the last words of the song they throw away the coco-nuts and the brands in the direction of the spirit-land, leaving the ghosts to wend their way thither, while they themselves return to the village.

Sir James George Frazer, *The Golden Bough*

I looked at the blurry image on Umbu's phone. It was a tiny pixelated snapshot of a photo, which was itself badly out of focus. We passed it round, all trying to decipher what the image was supposed to *be*. I could make out a vaguely conical, slightly curved white object. Umbu looked at me excitedly. 'It's a mili mongga tooth!' He'd put the word out through his network of friends and relatives that we were on an important quest to learn the truth about Sumba's mystery wildman, and someone had just texted him the photo. Apparently the tooth had been found in the 1950s by some people who'd excavated a mili mongga grave; and there were also 'big' bones in the grave. Umbu made some phone calls; apparently the tooth was now in Kupang, the capital of Indonesian West Timor. We'd hit a dead end – even Umbu's contacts didn't reach that far. However, maybe we could find out more about these big bones. 'The grave is in a village near Mangili,' Umbu said. So we piled into the car and headed out east.

It was very dry around Mangili. This south-eastern corner of Sumba was so isolated that until the 1970s the journey from Waingapu took about a week; we were lucky that it was now only a few hours' drive from Rindi. Eventually we arrived at the tiny village where we were going to stay. Umbu introduced us to a family he knew here already, as always. In front of the house was a tiny stone figure of Katuada, one of the Marapu deities. Other old stone figures stood in the paddock and down towards the sea.

It was late. We started setting up our bedrolls, sleeping bags and mosquito nets on the raised bamboo platform in front of the house. All of the adults from the neighbouring houses stood watching in silence as we awkwardly tried to undress in our sleeping bags without exposing anything particularly compromising. At the other end of the raised platform was a huge television screen broadcasting an Indonesian current affairs programme about American international relations. All of the village children sat in front of the blaring screen, engrossed in global geopolitics and ignoring us completely while their parents inspected us in intimate detail. The tension broke when Umbu asked if our *bulsaks* were comfortable – amidst our exhausted laughter, I learned that this was the Indonesian word for mattress rather than a part of the male anatomy.

Early the next day we all sat together drinking coffee, making plans in the morning's sharp light and long shadows. A pair of green pigeons, another of Sumba's endemic bird species, flew fast and low over the scrub at the edge of the dry paddock. The mili mongga grave was over in the old village, which our host tentatively reckoned was maybe 600 years old. Umbu looked at the pigeons through the clear morning air. 'The people here say that if a bird flies across the mili mongga's grave, it will die,' he mused.

It didn't take long to get to the old cemetery, an open expanse of long yellow savanna grass dotted with ancient twisted trees. Huge tombs sprouted from the knee-high grass like crumbling grey mushrooms. The sun rose in the sky; everything was hot and quiet.

The old man who had accompanied us from the new village explained that the cemetery was for local rulers and their families. The tombs certainly looked fit for a king; they were truly megalithic, with vast thick capstones supported on stout rocky legs. And the tomb in front of us – the tomb of the mili mongga – was similarly huge. Its capstone was a huge block of grey limestone, almost five metres long by three metres at its widest point, and one metre deep. It was raised up at each corner on four heavy stones, pitted and weathered by the elements. A casambi tree had grown up through the capstone and shattered the rock; thick vines hung down onto the tomb from its branches, and the stones and the ground around them were covered by dry brown leaves. The royal tombs had all been looted for gold over the years, but the mili mongga tomb had never been looted, our guide said, because it was bad luck. People never came to visit it. We crouched down amongst the shadows beside the tomb and peered underneath. There was no way that anybody could even squeeze under there to take an exploratory look.

Many generations ago there had been lots of mili monggas living in the jungle all around where we were standing, the old man continued. He pointed to the dry stone wall that marked the boundary of the cemetery, taller than a person and made of large weathered lumps of limestone. 'That wall was made by a mili mongga,' he explained. I looked to Umbu in surprise. The red handkerchief tied round his neck made him look like he was dressed for a proper adventure. 'Yes, sometimes the mili monggas used to help people,' he confirmed. 'In the forest near Rindi is a cave where a mili mongga lived two hundred years ago. She used to help people from Rindi with the ploughing.'

First I'd thought that mili monggas were just a bit of cryptozoological fun, then I'd wondered if the stories we kept hearing about wildmen might be based on vague memories of encounters with monkeys or discoveries of fossil bones. But imaginary yetis and stegodon tusks can't help with the DIY or the farming. Surely only other people could do that? Thinking back, the stories we'd heard at the start of our trip had also described mili monggas in ways that made them seem curiously human in some respects.

A new possibility seemed to be emerging. Might the stories about mili monggas just be stories about ... other people? After all, if these tales were based on some sort of reality, then a human reality was certainly more plausible and prosaic than a creature unknown to science. But this raised more questions. Why would these 'other' people be seen as so different from the locals? If mili monggas *were* people, what sort of people might they actually be?

People can become 'othered' in many ways. This can happen both to particular individuals and to wider social groupings such as ethnic groups. Such targets become perceived as somehow marginal or liminal, and often as outcasts or deviants. This categorisation is fundamentally dehumanising; as history has soberingly shown time and again, thinking of certain people as 'other' makes it easier not to treat them with respect and value. But surprisingly often, the very boundaries between what is considered human or non-human can also become blurred when thinking about other people. Could this help explain the stories of giant feral wildmen on Sumba? If certain people were once assigned to different cultural categories in the past, might these later become misremembered as biological categories? Or in some cases can people really be mistaken for non-humans?

Certain unusual medical or behavioural conditions, if encountered in rural communities, might potentially provide fuel for local stories about wildmen. Maybe the most striking is hypertrichosis, an extremely rare medical condition characterised by abnormal hair growth all over the body. People with this condition sometimes ended up in circuses or freak shows, and were advertised to the credulous public as 'missing links' between apes and humans – members of primitive tribes who still lived in trees or caves, thus providing living proof of Darwin's theory of evolution. Some, such as the nineteenth-century Mexican performer Julia Pastrana, were taxidermied after their death and continued to be exhibited around circuses and amusement parks.

From Romulus and Remus to Mowgli and Tarzan, many cultures also have tales of children being raised by animals. Numerous instances of 'feral children' have been documented around the world over the past few centuries. Such children spent long periods of time isolated from human contact and often experienced a combination of abandonment, abuse and mental illness (with the former often bringing about the latter), so lacked many basic social and behavioural skills. Local accounts reported these children being found in the wild with a range of foster animals, from wolves, bears, monkeys and sheep through to gazelles and even ostriches, among which they allegedly lived for months or years. Lucien Malson, in his classic account of 'wolf children', even described a 'pig-boy' called Clemens 'who had a particular passion for green plants and had to be restrained whenever he got near a cabbage patch.'[1] Whether or not such children have truly been raised by other species in the absence of normal human care, they often act superficially like animals due to their social deprivation, for example preferring to move around on all fours.

Ever since antiquity, feral children have fascinated scholars trying to understand the origins of language, what aspects of human nature were innate or learned, and the boundaries of what could be considered human. Lord Monboddo, the eighteenth-century thinker who drew attention to the similarities between humans and apes, was similarly fascinated by feral children, writing that the discovery in 1725 of 'Peter the Wild Boy' was more remarkable than the discovery of Uranus.[2,*] Recent reassessment of a contemporary painting of Peter suggests that in fact he had a rare genetic disorder called Pitt–Hopkins syndrome.[3] A couple of centuries later in the heady academic atmosphere of the 1960s, Malson used feral children to champion his Marxist-inspired philosophy that human behaviour is not inherited, and that 'the idea that man has no nature is now beyond dispute … before his encounter with others man is nothing but a notional quantity as thin and insubstantial as mist.'[4]

Sometimes people have also been mistaken for 'missing links' based on other aspects of their physical appearance. While visiting rural Wales to research the ancestry of different human populations across Britain, the geneticist Bryan Sykes was told a story – by a man at the bar of the local hotel – about two brothers who had lived in the hills

---

* An anonymous contemporary author, possibly Jonathan Swift, also described Peter as 'the most wonderful wonder that ever appeared'.

above the Welsh market town of Tregaron. Because of their heavy features, this pair of bachelors was widely thought to be a pair of surviving Neanderthals. In an example of charming hospitality, during the 1950s and 1960s the brothers would allegedly host annual visits for local schoolchildren, and provide fizzy pop and cakes while the teacher used them as live exhibits in lessons on human evolution.[5,*]

And, of course, real giants do exist. People who produce unusual amounts of growth hormone (a condition called acromegaly, caused by a benign tumour on the pituitary gland) can reach heights of almost nine feet, although they often die relatively young due to the extra strains on the body that accompany supernormal growth. (Note that this is a very different mechanism to how animals evolve slowly into giant forms on islands, whereby slightly larger-bodied individuals tend to preferentially survive and reproduce over time in response to natural selection across many generations.) In the past, giant people – such as the seventeenth-century John Middleton or 'Childe of Hale', or the eighteenth-century 'Irish Giant' Patrick Cotter, who became an attraction in Bristol – were inevitably a source of fascination and wonder. As G. Frankcom and J. H. Musgrave noted in their account of Cotter, the populace of the time 'readily paid to see any deviation from the human norm, whether tall, short, fat, deformed or even spotted.'[6] The Blaise Castle House museum in Bristol still possesses some of Cotter's personal effects, including a cast of his hand, and I was entranced by his huge shoe when I visited as a child.

Entire categories or groupings of people have also often been interpreted or labelled as non-human, sub-human or animal, because of their different and unfamiliar appearance, customs and origins, or because they subvert social rules or don't fit into accepted social frameworks. Such interpretations can be made about particular groups of people by other ethnic groups or cultures, or by other members of their own society; and people have been diminished by others as 'unpersons' long before George Orwell coined the term in *Nineteen Eighty-Four*. For instance, outlaws in medieval England were referred

---

*James, one of the brothers, is also illustrated in H. J. Fleure's *A Natural History of Man in Britain* (Collins, 1951), where he is described as 'showing features akin to those of man of the later Palaeolithic age'.

to as 'wolf's heads' or *caput lupinum*, a symbolically animal attribute indicating they had lost their spiritual right to be considered human as a result of heinous, 'bestial' crimes; they would die in the wild spaces without last rites or burial, and could legally be killed without penalty. Similar wolfish qualities were formerly associated with outlaws who skulked dangerously at the margins of society in other parts of medieval Europe such as Iceland. Gregory Forth recounts a comparable example from Buru, an island north of Flores, where a nineteenth-century account of hairy human-like creatures supposedly called *gibar bohot* turned out to refer to fugitives from other islands who hid in caves and stole from fields. *Gibar bohot* just means 'bad person' in the local language. Eventually the locals retaliated and killed the fugitives; subsequent examination of their beheaded remains by a western visitor confirmed they were definitely human.[7] Indeed, some yeti sightings are also suggested to be glimpses of escaped convicts hiding in the mountains to evade the Tibetan police.[8]

More widely, many of the monstrous races in old bestiaries were probably inspired by travellers' tales and hearsay, or encounters with unexpectedly 'different' ethnic groups. Early European explorers justified their conquest of the native people of the Caribbean by describing them as savage cannibals (the words 'Caribbean' and 'cannibal' share the same origin), which probably inspired Shakespeare's similarly-named island inhabitant Caliban in *The Tempest*, an inhuman 'mooncalf' and monstrous son of a witch. Other European visitors to far-flung parts of Asia often reported seeing or hearing about people with unusual characteristics, such as a tribe from the Malay Archipelago who were supposedly only active at night. In the words of the seventeenth-century traveller François Leguat:

> … *what is most Remarkable in them, is that their Eyes cannot endure the Light, and they always see best a-nights, so that they turn Night into Day, and Day into Night. I have often met of them trudging along with their Eyes almost shut, because they were not able to look on the Light.*[9]

Other explorers visiting the islands of Southeast Asia, such as Antonio Pigafetta (one of the few survivors of Magellan's circumnavigation of the world), reported people covered in hair and who 'eat only raw human hearts with the juice of oranges and lemons'.[10] The eighteenth-century Swedish traveller Christopher Braad, whose writings influenced Monboddo and Linnaeus, visited Malacca and described 'a kind of wild humans (if they may be called human)' with 'short hair, which they are said to tear off to facilitate their moving among the trees in which they have their huts.'[11] As late as 1884, Robert Sterndale's *Natural History of the Mammalia of India and Ceylon* reported some supposed 'Monkey-men' encountered in the jungle by a Mr. Piddington, who were 'very short in stature, with disproportionately long arms, which in the man were covered with a reddish-brown hair.'[12]

Uncertainty over whether to classify newly-encountered peoples as people at all was further influenced by theology, which played a role in biasing perceptions about anthropology as well as palaeontology. For example, the discovery of the Aztec civilisation by conquistadors generated intense debate in Spain during the sixteenth century, as the Aztecs were not mentioned in the Bible. How could they be descended from Adam and Eve if people had not previously been able to cross the Atlantic? (An analogous argument was employed centuries later by anthropologists unfamiliar with island biogeography theory, to challenge the suggestion that *Homo floresiensis* was a genuine island-endemic hominin species.) Had God created a second Adam and Eve for the New World, or were these 'half-men' not true people at all? Even the astonishing artefacts, ornaments and grand cities encountered by the conquistadors were not sufficient to convince commentators in Europe of the Aztecs' humanity: 'this is no proof of human cleverness, for we can observe animals, birds, and spiders making certain structures which no human accomplishment can competently imitate.'[13]

Similar confusion – whether from ignorance, prejudice or a biased worldview – is also seen in other cultures. Ancient Chinese texts such as the *Classic of Mountains and Seas* describe numerous hairy ape-like beings with curiously human behaviours such as singing and the ability to make wine. It is unclear whether these are distorted accounts of gibbons or other primates then found in southern China, or alternately of the various human ethnic groups who also lived there.[14] Other types of Chinese wild people, with names such as 'mountain uncle' and 'mountain auntie', were described as very similar to other people in appearance and behaviour – but, worryingly, could legitimately be eaten.[15] In an analogous manner to European intellectual debates over how feral children and travellers' tales might define what it meant to be human, the ancient Chinese worldview regarded 'the alien groups living outside the pale of Confucian society as distant savages hovering on the edge of bestiality', in the words of sinologist Frank Dikötter.[16] As late as the 1930s, even the Chinese characters for many of these ethnic groups were partly derived from the written characters for animals, and contained the radicals or character classifiers for various different species: the Di were associated with dogs, the Qiang with sheep, and the Man and Min with reptiles.[17] Even modern traditions of China's own wildman, the yeren, are suggested to have originated in historical encounters with bearded Europeans such as Greek traders from the Qin Dynasty onward.[18]

The 'dominant' powers throughout history – the colonisers, conquerors and intruders into other peoples' domains – have thus often 'othered' the various cultures they encountered, regarding them as no different from animals and sometimes even as yeti-like hairy wildmen. Demonising Indigenous cultures and portraying them as savages also made it more morally acceptable to subjugate them and take their land. Conversely, first contact with European explorers was seemingly often framed in more supernatural terms by Indigenous peoples. Precisely how most of these contact events were interpreted by different cultures around the world is now lost to history, as the peoples who were encountered during the period of European exploration and expansion were typically non-literate and left no written record of the event. As we have seen, tragically many of these cultures were also rapidly disrupted or destroyed through exposure to new European diseases. All we have in most cases – if we're lucky – are the biased historical accounts of European chroniclers who tried to make sense of unfamiliar cultures in the absence of a common language, or indeed any real frame of reference for the usually mystifying behaviours they witnessed. For instance, both the Spanish conquistadors and Captain Cook are often said to have been mistaken for returning gods by the cultures they contacted. The Aztecs supposedly thought Hernán Cortés was the feathered serpent deity

Quetzalcoatl, whereas the Hawaiians mistook Cook for Lono, a god of peace and fertility – first honouring him, then killing him when they realised he was human after all. However, in both cases the reality was probably either more complex and nuanced, or in fact something else entirely was going on that was lost on contemporary European observers.[19]

Only in the case of more recent contact events can we begin to understand what was going through the minds of Indigenous peoples when they first encountered Europeans – because it was possible for anthropologists to go back and actually interview the specific individuals who were there at the time. The best-documented contact event took place in the 1930s, when a group of Australian explorers led by Michael Leahy, who were prospecting for gold in the unexplored highlands of New Guinea – generally thought at the time to be uninhabited – instead found an 'untouched' agricultural society of almost a million people. The Australians were widely thought to be the reincarnated spirits of dead relatives or ancestors – in part because the explorers had arrived from the east, the direction where local traditions believed that people went after they died. Even the expedition's New Guinean porters were thought to be dead clansmen. Other specific attributes of the new arrivals, such as the false teeth of one of the explorers, were taken as further proof that these people must be dead. Their white tents were thought to be made from clouds, and their lanterns must be pieces of the moon that were brought from heaven. The highlanders also thought that the all-male party of explorers must have women hidden inside their rucksacks.[20]

The fact that dead relatives had unexpectedly returned prompted many different reactions. Many of the old people became emotional and wanted to embrace the explorers, who had to set up a rope fence to keep them away.[21] But the highlanders were not naïvely credulous – they wanted proof to confirm that their visitors were indeed dead people rather than just strange living people after all. The dead were supposed to adopt human form by day but would revert to skeletons at night – but the rope fence made it difficult to see inside the Australians' tents to confirm whether such a transformation took place each evening. Other lines of logic were explored. Because the strangers wore tight-fitting clothes, they must be concealing huge penises wrapped around their bodies – and it was hard to believe they could contain any excrement. Eventually one of the villagers managed to hide and watch an Australian visiting a screened-off latrine pit, and learned that 'Their skin might be different, but their shit smells bad like ours.'[22] The biggest shock came when they spied on the strangers washing themselves with soap – the foam seemed to resemble the milky pus of a corpse. 'Our minds were in a turmoil when we saw such things!'[23] Some highland communities encountered by Leahy's expedition later told anthropologists they realised quite quickly that the strangers were not dead people, but others took years to comprehend it.

The New Guinean highlanders were not alone in connecting strange western visitors to beliefs in supernatural beings. Colonial encounters with Melanesian peoples across the south-west Pacific led to the emergence of numerous so-called 'cargo cults' such as the John Frum cult on Tanna, which are based around spiritual entities seemingly inspired by westerners such as World War II servicemen. There is even one that promotes the divinity of Prince Philip. These sociologically complex religious sects appear to serve many purposes, including to provide frameworks for cultural revitalisation and self-determination, establishing new social relationships with outsiders, and addressing social change resulting from colonialism. And sometimes western visitors have been perceived in supernatural terms as entities surprisingly similar to mili monggas. For instance, naked hairy giants called *biri-biri* and *yawas*, which supposedly inhabit the jungles of the Wallacean island of Halmahera, are said to be degenerate surviving seventeenth-century Portuguese colonists.[24]

We cannot now reconstruct the cultural fall-out of such encounters around the world from further back in time, such as thousands of years ago when technological changes prompted communities of farmers to move around Europe and repeatedly cross the channel into Britain. We will probably never know what happened when these new migrant tribes encountered people who already lived in the land they wanted to colonise. However, it's been speculated that these events might also have been framed as supernatural encounters – either by people at the time, or as the tales of such momentous meetings were passed down the generations and slowly reshaped by each re-telling. Variants of this hypothesis were widely discussed amongst folklorists and antiquarians throughout the nineteenth and early twentieth centuries. Canon John MacCulloch noted that 'where a territory has been conquered, the aborigines are apt to be regarded in course of time as having a kind of spirit form'.[25] Herbert Fleure, Professor of Anthropology and Geography at the University College of Wales, proposed that widespread British legends of 'fairies or moorland folk to whom iron was taboo but who were skilled in herbal medicines' could in fact represent ancient tales about the prehistoric Neolithic or Bronze Age inhabitants of Britain, recounted by the descendants of the Iron Age peoples who assimilated or displaced them from the landscape.[26]

Or might fairies be based instead upon tales of ancient Irish invaders, or valiant heroes whose identities and origins are now long-forgotten?[27] It was even speculated that a pre-Celtic race of tiny people might have existed in Britain during prehistory (the 'pygmy-Pict' theory). Walter Scott and others thought that fairies, mermaids and selkies were all distorted memories of encounters with supposedly tiny Finnish or Lappish people, living in underground mounds or paddling about in kayaks[28] (although how Finns had been encountered by ancient Britons was left unclear). The idea that fairies and other supernatural entities might be based upon ancient historical reality also appealed to horror story writers such as Arthur Machen, who made it a repeated theme in his strange fiction, and considered that rural traditions about the Little People in places such as the Forest of Dean were probable folk memories of pre-Celtic inhabitants. Machen proposed that 'There was such a race of short, dark people, who did, in fact, live subterraneously, and survived far into the Celtic age. But the people of this race must have been about four feet—not fourteen inches—high.'[29] These ideas (at least some of which were made seriously) are untestable, but intriguing. At least they're more convincing than the proposal made by James Cririe in 1803, who suggested that accounts of fairies were inspired by 'Druids or aborigines taking refuge in subterranean dwellings and emerging by night to exercise their limbs by dancing' (this idea was mocked as a 'marvellously absurd supposition' even at the time[30]). They're also broadly similar to tales about strange supernatural people in other parts of the world, such as the Vazimba of Madagascar, who are suggested by some anthropologists to be possible distorted folk memories of an earlier wave of human settlers.

And there's even direct scientific evidence that people with differing ethnic backgrounds might have been viewed as inhuman wildmen. After his involvement with the Tregaron Neanderthals, Bryan Sykes looked into a mysterious historical tale from Abkhazia, a breakaway former Soviet republic in the mountainous Caucasus region. In the 1850s, what was described as a huge female wildman covered in long reddish-brown hair was caught in the forest and held captive until her death in 1890. Nicknamed Zana, she was reportedly feral, unable to speak, and refused to wear clothes, although she eventually performed menial tasks such as grinding corn in a watermill − basic physical work very similar to the jobs that mili monggas are supposed to have done in Sumba. She also had several children with local men. Cryptozoologists such as Bernard Heuvelmans were convinced that Zana was a late-surviving Neanderthal − a perfect example of their 'prehistoric survivor' hypothesis. However, Sykes conducted genetic analysis on tooth and saliva samples from Zana's descendants, and found that she was in fact fully human − but was originally from sub-Saharan Africa rather than the Caucasus.[31] How she ended up living a tragic life so far from home remains a mystery; but her story is a cautionary one, demonstrating how fantastical stories of supposed wildmen and other bogeys might actually be grounded in local ignorance of other races, with initial witness accounts becoming flavoured with later embellishment and exaggeration.

So far our time on Sumba had focused on trying to reconstruct the island's ancient history for the first time. Our new fossil discoveries had revealed the former existence of now-extinct giant rats and unusual monitor lizards. But other than these finds, only the dwarf stegodon from Watumbaka was known from the island's recent fossil record. We thus had no idea whether there had also been a strange endemic human species similar to *Homo floresiensis* here to be encountered by modern humans when they spread across Wallacea 50,000 years ago. However, consideration of Sumba's more recent history – recent in the sense of what had happened during past centuries or millennia, rather than tens or hundreds of thousands of years ago – might also be informative. Was there any evidence of local encounters between different peoples that might account for the strange stories about mili monggas? As it turns out, even this remote out-of-the-way island has been exposed to a surprising diversity of different groups and types of people throughout its history.

Sumba, like the many other islands across Southeast Asia, experienced multiple successive waves of colonisation by different groups of modern humans. As we've seen, the first dispersal of *Homo sapiens* through Wallacea was part of the incredible prehistoric overwater migration of humans all the way to Sahul, the Ice Age continental landmass comprising modern-day New Guinea, Australia and Tasmania. We still have much to learn about this first prehistoric expansion of modern humans out of Asia. Recent research involving ancient DNA analysis of a Holocene hunter-gatherer skeleton from Sulawesi has revealed that this individual was related to modern-day Indigenous Australian and Papuan groups, but was part of a previously-unknown divergent human genetic lineage.[32] These colonists still used relatively basic stone tool technology characteristic of the Palaeolithic, but were culturally sophisticated in other ways; in addition to their rich cave art and ability to catch oceanic fish such as tuna, they also showed considerable further technical skills. This is ably demonstrated by the recent discovery of a 31,000-year-old skeleton of an adolescent from Borneo, who had the lower part of their left leg amputated and survived for up to nine years afterwards. This remarkable find represents by far the earliest evidence of successful surgical limb amputation in the archaeological record.[33]

Unlike the Indigenous peoples of Australia and New Guinea, the people of Sumba cannot trace their ancestry easily back to this ancient human migration. Instead, they're the descendants of the more recent 'Chinese steamroller' event, when Austronesian language-speaking Neolithic farmers spread out from Taiwan a few thousand years ago, first southward across the islands of the Sunda Shelf, the Philippines and Wallacea, then eventually across all the inhabitable islands of the tropical Pacific. This remarkable human migration is evidenced by the arrival of distinctive new types of pottery and other so-called 'material culture' in the late Holocene archaeological record across Wallacea and beyond. The intrepid Austronesian argonauts made it as far as Easter Island, one of the most remote inhabited islands in the world, and eventually reached New Zealand about 700 years ago. As with other prehistoric human migrations, the reasons for this spectacular diaspora are unclear, but it might have been triggered by the invention of what Jared Diamond has called the 'speedboat to Polynesia' – the outrigger canoe, where two logs or 'outriggers' are lashed onto poles on either side of a dugout canoe. This major technical innovation prevents the canoe from capsizing even in stormy seas, and allowed the Austronesians to break free from the coast and set out into the open ocean as a maritime culture. In the words of author Christina Thompson, the speed and extent of this population expansion has made the Polynesians, the easternmost ethnolinguistic subgroup of Austronesians, the world's 'most closely related and the most widely dispersed people.'[34]

As they spread throughout Southeast Asia, the Austronesian colonists would not have found a series of uninhabited islands. Instead, most or all of the habitable landmasses they reached were already home to existing populations of settlers, descendants of the previous prehistoric human migration. It was these first hunter–gatherer colonists who presumably wiped out the indigenous hobbits on Flores and the dwarf stegodons on many Wallacean islands – and then, without the ability to terraform their landscapes through agriculture, became culturally adapted to the unique conditions of the islands on which they found themselves. It's suggested that the pre-agricultural hunter–gatherer landscape on the arid island of Sumba 'probably resembled aboriginal Australia, where human population density scaled with rainfall'.[35] As we've seen, some descendants of these first Southeast Asian colonists – the Indigenous hunter–gatherer Qata peoples, such as the Aeta and Mamanwa of the Philippines – still hang on today as marginalised communities on a few Southeast Asian islands. However, the handful of Qata peoples that still exist are in decline due to competition for land and resources with their Austronesian neighbours, exacerbated by a depressing litany of socioeconomic stresses including high levels of disease, malnutrition, alcohol abuse, and endemic violence. Nearly all of these uniquely distinct cultures have already long since vanished.

Other than these insights from current-day interactions between Qata and non-Qata communities, it is challenging to reconstruct exactly how the Austronesians replaced most of the original Palaeolithic inhabitants of Southeast Asia thousands of years ago – whether through violent rapid displacement, or more gradual and potentially more peaceful assimilation of genes and cultures. Interestingly, some current-day inhabitants of the Lesser Sunda Islands, such as the people of Baucau and Viqueque in East Timor, are generally similar to their Austronesian neighbours but speak languages closely related to those of non-Austronesian inhabitants of New Guinea. So-called Papuan languages were spoken by otherwise culturally Austronesian peoples as far west as Sumbawa during relatively recent history, until that island's Papuan speakers were wiped out by the catastrophic eruption of Mount Tambora in 1815. Several other languages that are classified as Austronesian, including those of Sumba, actually have very atypical grammars and vocabularies, suggesting that these might be derived from ancient Papuan languages too. Indeed, a comprehensive study of languages spoken across Sumba found that 65 per cent of sampled words could not be traced back to proto-Austronesian, 'and may have been absorbed from a now extinct indigenous source'.[36] This relatively widespread regional survival of linguistic bits and pieces that probably date from before the 'Chinese steamroller' hints that maybe Austronesian colonisation and cultural domination was a relatively gradual process.

Recent genetic studies of modern human populations on islands such as Timor have further revealed that substantial genetic mixing took place over time between the two different colonising groups.[37] The distribution of Y chromosome gene lineages characteristic of either Austronesian or Papuan peoples that are seen in human populations across Sumba also suggests the initial spread of Austronesian colonists 'was accompanied by frequent intermarriage with resident hunter-gatherers'.[38] However, such studies have conversely found no evidence for earlier admixture between modern humans on Flores and *Homo floresiensis*; this has been specifically investigated for the Rampasasa pygmies, which turn out to be an evolutionarily recent instance of population dwarfing, completely independent of any ancient hobbit genes.[39] The first overwater migration of *Homo sapiens* across the archipelagos of Southeast Asia thus seems to have simply swept away the region's oldest island-adapted human species, who had probably been there for over a million years.

It's therefore possible that some now-vanished Qata peoples might have coexisted as culturally and ethnically distinct groups alongside Austronesian communities for considerable lengths of time − as a hypothetical example, as tribes somewhere on Sumba. If so, based upon other contact scenarios documented from around the world, it's not impossible to imagine that such peoples might have been seen as something really very different. Indeed, some Qata peoples have historically been perceived by neighbouring communities to possess supernatural or non-human attributes. The coastal Malays used to believe the Orang Asli, Malaysia's 'Original People', possessed magical powers,[40] whereas the Maniq of southern Thailand were sometimes viewed as 'fire-making monkeys'[41] and appear to have been locally associated with the 'monkey followers' of Rama in the *Ramayana*.[42] Early Dutch ethnographers suggested that various local Indonesian traditions of human-like 'wildmen' might actually be based upon aboriginal peoples.[43] And a remarkable recent archaeological discovery from Taiwan provides a further dimension to understanding these relationships. Small-bodied Qata people are not known to have existed historically in Taiwan, the cultural starting point for the 'Chinese steamroller' demographic expansion event, but many of the island's Austronesian ethnic groups have legends of 'tiny dark people'; some of these legends refer to these tiny people as their enemies, and the Taiwanese Saisiyat culture practices ritual worship of black pygmy spirits. Then a fascinating study in 2022 reported the 6000-year-old skeleton of a small-statured hunter-gatherer individual from the Xiaoma Caves in eastern Taiwan, which ancient DNA analysis revealed was most closely related to living Qata populations from the Philippines and Andaman Islands. This extirpated Qata population (which is now speculated to have survived into recent centuries) has been interpreted as the likely origin of the widespread Taiwanese legends of Qata-like 'entities'.[44]

So might this begin to explain tales of mili monggas on Sumba? Do these giant wildmen represent a cultural memory of the former Qata inhabitants of the island; if so, was it the case that 'antiquity magnified them, as the mists of the Brocken magnify men into enormous spectres'?[45] Alexander Haggerty Krappe, writing in *The Science of Folk-Lore* in 1930, noted that:

> ... *more than once giants were called after a foreign people, usually neighbours or former inhabitants of the country more feared than liked. Thus, for example, the German word* Hüne, *denoting a giant, is unquestionably a derivation from* Hun, *which word designated not only the subjects of Attila but also the mediaeval Hungarians.*[46]

Or is this all just more speculation?

Maybe not. Fascinatingly, anthropologist Janet Hoskins, who conducted extensive fieldwork in rural communities in West Sumba in the 1970s and 1980s, provides some tantalising indications that distorted local memories of Sumba's original Qata population might persist in local folklore. She reported that most West Sumbanese thought their ancestors had reached Sumba by crossing over a stone bridge that used to link the island with lands to the west, but a handful of local people were instead supposedly descended from earlier inhabitants called the Lombo or Karendi people. These people were said to have no knowledge of fire or agriculture, which were both introduced by the new people who arrived across the stone bridge. The island's original inhabitants could also supposedly 'assume the form of wild animals or carrion-eating witches'; they were no longer seen as quite 'human'.[47]

Maybe thinking about the specific jobs that mili monggas supposedly did for local people can also provide further insights into what might be going on. They are supposed to have built huge stone walls. In fact, stories of monstrous beings constructing walls or other monumental stone structures in the distant past are widespread. On nearby Flores, a race of small but extremely strong beings called the ana ula, part of the folklore of the Poma people who live to the north-west of the Nage, are said to have helped the Poma's ancestors build defensive stone ramparts.[48] As an interesting aside, the ana ula also apparently love to steal aubergines, their favourite food. Elsewhere, dry stone walls made of massive unworked boulders in ancient Mycenae are termed 'Cyclopean masonry', because the classical Greeks believed that such huge constructions could only have been built by the giant Cyclopes – the tales of which were possibly inspired by fossilised skulls of tiny elephants similar to the stegodons once found on Sumba. Horror writer H. P. Lovecraft also repeatedly attributed the construction of Cyclopean architecture to his pantheon of invented ghastly ancient deities, the Great Old Ones and Elder Things (fun fact: Lovecraft uses the word 'cyclopean' a grand total of 47 times in his weird fiction[49]).

Prehistoric megalithic structures and other huge stone constructions across Europe, from Britain's stone circles to the Bronze Age nuraghes of Sardinia,[50] were often attributed to the work of giants in past centuries. Stonehenge was once called Chorea Gigantum or the Giant's Dance; the megalithic tombs of Cornwall are known as quoits because of the old story that giants used to play games with their capstones; and the decaying ruins of the Roman city of Bath were described by a Saxon observer as 'the work of giants'. Centuries later, upon seeing a tower, Kaspar Hauser – the most famous of the feral children who were the talk of Enlightenment Europe (although he may have been a fraud) – supposedly said 'the man who built it must have been very tall indeed.'[51] Alexander Krappe considered that 'whenever a population finds in a newly-settled country huge monuments built by the former inhabitants, it almost automatically attributes those ruins to supernatural beings, giants, fairies, the Devil, or, in the Orient, the jinn', and 'to the Rumelian peasants the Hellenic structures are the works of giants'.[52] This common misconception that huge prehistoric stone structures could only have been built by huge prehistoric people is the intellectual precursor to the pseudoscientific 'ancient astronaut' theories popular during the 1960s and 1970s (which posited that 'primitive' peoples couldn't have built anything impressive themselves, but instead must have been helped by aliens). Indeed, 'the wild man as helper' represents an entire category of tales within the widely-used folktale classification system developed by folklorists Antti Aarne, Stith Thompson and Hans-Jörg Uther, and includes such stories as the German fairy tale of the hairy wildman 'Iron Hans' that formed part of the collection of the Brothers Grimm.

In fact, what we know of Sumbanese history tells a different story. The stone walls that encircle ancient villages across the island are thought to date back to the time when European and Asian trading powers began to take notice of this far-flung corner of the Malay Archipelago. From the sixteenth century onward, ships from Spain, Portugal, England, China and the Middle East, and later from the Dutch East India Company, all began stopping off at Sumba to obtain the island's sandalwood, tiny horses, honey, wax and pepper, which they exchanged for precious metal. The island's importance for international trade in sandalwood led to Sumba becoming known as Sandalwood Island. This was its recorded name as early as 1522, when Antonio Pigafetta reported sailing past an island known locally as Cendana ('sandalwood' in Malay and Indonesian), and a Dutch ship was wrecked upon 'the unknown island of Sandalwood' in 1636.[53] Local sandalwood was used to make fragrant incense and clothes chests,[54] and is still highly prized for its aromatic oil, making it one of the most expensive and sought-after woods in the world. But the metal weapons that became common on Sumba as a result of this trade led to an intensification of local warfare, which had always been bubbling away across the island and now became more widespread and lethal, as tribal rulers sought to consolidate and expand their power-bases with the help of this new technology. Prisoners acquired from these skirmishes were referred to as 'feet of wild pigs, paddy gathered on horseback',[55] and were typically sacrificed. The extensive construction of huge walls formed part of a system of strengthened local defences to protect vulnerable villages. Similar unwitting disruption to local cultures as a consequence of European trade in metal weapons also occurred elsewhere across the Austronesian world. The arrival of guns in New Zealand following contact with Europeans led to a period of vicious intertribal conflict known as the Musket Wars, which resulted in tens of thousands of deaths, and during which a group of Māori sailed to the nearby Chatham Islands and exterminated nearly all the Indigenous Moriori people.

Even greater social disruption was brought about on Sumba during this period by a growing slave trade across the Malay Archipelago. The dehumanising process of slavery is the starkest historical example of equating people with animals. The earliest European visitors to eastern Indonesia reported observing a local trade in slaves, and Sumbanese culture itself was composed of a rigid caste system of hereditary classes, including a slave class, the *ata*, people who could be bought and sold. Slaves were also sacrificed as offerings to the spirits, or to cover sacred drums with human skin.[56] However, the newly-forming colonial empires needed manual labour for their plantations in Asia, the Indian Ocean territories and South Africa, and the disruption of local politics by European and Arab trading and occupying powers had knock-on effects across the region. In particular, there was mass emigration from southern Sulawesi of sea-going Makassarese people, many of whom settled along the coast of Flores and became fierce slavers. Sumba became a hotspot for slave raids, with raiders making incursions deep into the island; the slavers brought with them new waves of smallpox,[57] and traded more muskets and gunpowder for slaves, thus further increasing local political tensions. By the nineteenth century, slave traders were also visiting Sumba from Bali, Lombok, Sulawesi, Buton, Sumbawa and Komodo – and Dutch, English, French, Chinese and Arab traders were also all involved. By now the island's north-west coast had been largely depopulated by slave raids, and even shipwrecked Europeans were captured into slavery.[58]

The local slave trade was still widespread at the end of the nineteenth century, when a contemporary observer wrote that 'everywhere on Sumba one finds the remains of settlements that have been laid waste and the inhabitants of which have been murdered or taken away as slaves',[59] and 'it often happened that an isolated Sumbanese in strange territory was taken captive'.[60] Conversely, many of Sumba's local rulers also imported slaves from elsewhere across Indonesia, probably including different ethnic groups such as New Guineans, who are known to have been traded across the region. Around this time, the first permanent Dutch colonial officer stationed on Sumba reported that constant warfare and slaving had 'reduced the value of a human life to a very low level ... often well below that of a horse.'[61] The local slave trade (both export and import) was only stamped out when the Dutch placed the island under military control in 1906. Rodney Needham, who conducted anthropological research in Sumba in the 1950s, reported the cultural scars of historical slavery remained evident in rural communities even many decades later; 'the appearance of a strange vessel out at sea, or just the rumour of one, would provoke all the signs of a general panic; men looked fiercely serious, and screaming women dashed to pick up their children.'[62]

So instead of being built by big hairy wildmen, the defensive walls across Sumba had been constructed during a period of social upheaval, danger and suspicion – when locals were 'fearful and mistrustful'[63] of strangers, the island was stalked by people from overseas trying to snatch villagers from their homes, and manual labour (such as wall-building and ploughing) was often carried out by imported slaves from different ethnic groups and with different languages. All this confusion continued until little more than a century ago, around the time when the mili mongga stories we'd been told at the start of our trip were supposed to have happened. It's not too much of a stretch to imagine that rural stories of encounters with not-quite-human beings – both on Sumba and elsewhere across Indonesia – might actually represent half-forgotten garbled or apocryphal accounts of what was originally some kind of very human encounter. Or – given the similarities between some of the story themes we'd heard about wildmen from Sumba, compared to stories recorded from other Indonesian islands – had this ebb and flow of different peoples also brought myths and legends from overseas, which then became assimilated into local folklore? Who knows; but it was possible.

Back in Waingapu, Umbu organised a series of meetings with local officials and other useful contacts. Everyone seemed happy with our project and excited by our discoveries so far. And, it seemed, everyone we met had a story to tell. A conspiratorial old man in the tourism department leaned over and told me about a sacred cave near Rindi that was full of gold, but that you could only see if you were very lucky. Then he launched into a long rambling local legend about a princess who wanted to be carried around by four men, so, in some sort of seemingly apocryphal lesson, her attendants cut their balls off and made a hill of testicles that she had to climb up and sit on by herself. 'No, no!' another official interrupted, taking a long drag on his cigarette. 'That's not the story at all. She was carried around by four eunuchs! They didn't have any balls at all!' he cackled. 'What happened was, whenever she needed to take a shit they always carried her to the same place, and eventually a mountain of shit formed at the spot. Then two different men from a neighbouring island both came to ask for the princess's hand in marriage. She had a gold belt, so the suitor who was able to throw the belt over the mountain of shit got to marry her.' These local fairy tales seemed more earthy than those of the Brothers Grimm, even in their original unsanitised versions. It was like Homer writing for *Viz* magazine.

A local journalist we visited also had an interest in Sumbanese folklore. He'd heard of somebody meeting a mili mongga once in East Sumba, and again we were told the story of someone else who was descended from a mili mongga many generations ago, the details of which had to be kept secret. In East Sumba, mili monggas were stupid giant cannibals that could speak only broken language, like a child's speech, and were scared of barking dogs and the sounds made by tokay geckos.

The journalist told us a story to show just how stupid mili monggas were. 'A mili mongga once dug up all of a farmer's sweet potatoes. It ate the small sweet potatoes, and hid the big ones behind its back while it ate. Some kids crept up behind it and stole the big potatoes with sticks, and when the mili mongga reached back to eat them too, they were gone. First it asked its fingers, "Do you know where the sweet potatoes are?" "No, we helped you to dig them up," replied the fingers. Next it asked its eyes, but its eyes answered that they had helped the mili mongga to see where to dig. Then it asked its back, its arms and its legs, but these other body parts all gave similar explanations about how they'd also helped gather the sweet potatoes. Finally it asked its asshole, but the asshole didn't say anything. This made the mili mongga think that the asshole must have stolen the sweet potatoes. So, it got really angry, and stabbed itself in the ass with a sharp stick and died.' I burst out laughing; I wasn't sure if this story added much to the body of mili mongga folklore we'd already collected, although I was very glad I'd heard it. But the journalist looked serious. 'The mili mongga had just eaten the children's mother, and they'd found out because it had lured them into her house and served them her breasts on a plate.'

Umbu also introduced us to a local minister who was supposed to know more about the mili mongga tomb we'd visited and the alleged giant's tooth found there. The incongruous mop of shiny black nylon hair perched at an unnatural angle on top of the minister's head did not mask his advanced years. The tooth had been obtained from the grave by his father, he said. Four generations earlier – seemingly in the nineteenth century, during the period of island-wide slave-trading and local fighting – there were two warring clans living near the village, but one clan always managed to defeat the other because it had two mili monggas fighting for it. The other clan then prayed to the Marapu spirits for bees, which duly arrived, and somehow killed the mili monggas and helped the underdogs win. As the story continued, the minister explained that one of the mili monggas died when a child hit it in the foot with a small spear, allowing the child's father to deliver a fatal blow. A further element of confusion was raised – it's possible we hadn't been shown the real mili mongga tomb, as the minister thought that both dead mili monggas had actually been buried on either side of the main gate to the graveyard to guard the village – but anyway, in 1964 a researcher from Java had apparently excavated the site and found a giant skeleton.

The minister's father had been present during the excavation and witnessed the huge size of the skeleton's leg bones, which were 'old yellow' in colour and looked like big bamboo. His father had ended up acquiring one of the teeth from the grave (it had apparently 'fallen out'; we didn't press the point further). The minister explained that mili monggas had once been common all over Sumba, but had hidden in the forests when people first arrived on the island. They were seven feet tall with huge heads, and covered in reddish-black hair like a westerner's beard. Mili monggas liked to eat snails and villagers would trade snails with them, but they were also cannibals – at least, they ate people, whether or not they were human themselves – and to have a mili mongga on your side in battle pretty much guaranteed victory. The name 'meu rumba' meant the same as mili mongga, but bad people were sometimes referred to as meu rumbas or 'wild cats' because their behaviour had made them just like animals; so this alternative name possibly just reflected the wildman's wildness rather than anything inherently cat-like about its appearance.

Was this a tale about a local yeti? Was it a folk memory of an Indigenous ethnic group present on Sumba before the arrival of Austronesians, or a clue that stories about wildmen were really about criminals or outlaws? All these conceivable explanations seemed equally possible from the things we were hearing. And the warfare connection was particularly intriguing. Throughout Sumba's endless historical conflicts, 'half of the art of making war lay in creating a fearsome appearance', with war leaders often heavily decorated and costumed and variously described as a large boar, the shining snail shell, the seed of a great tree – or 'a giant monkey'.[64] Were stories of giant man-beasts that could win wars actually part of this propaganda; were they really just local memories of victorious chiefs in battle-gear?

But how could a visiting academic really have dug up the grave of a giant without it becoming global news? The minister was adamant that this had happened, though. Then he produced a book written in Indonesian. He flicked through the pages, and showed us a better-quality version of the photo of the supposed mili mongga tooth that Umbu had on his phone. It was possible to make out more detail than before. It was the tooth of a sperm whale.

In the same way that fossils often inspired beliefs in monstrous creatures, unusual or oversized remains of modern-day animals have sometimes also become the basis for legends of fantastic beasts. For instance, the medieval English tale of the giant 'Dun Cow' is associated with relics that are actually old whale ribs and narwhal tusks. And all around the world, people have sometimes deliberately buried animal remains alongside, or even instead of, human remains. These practices were carried out for many reasons, including prestige, as sacrificial or apotropaic offerings, to provide companions or food for the afterlife, and other possible symbolic associations and religious or cultural reasons now lost in the past. Sea eagles were interred alongside people in Neolithic chambered tombs on Orkney; exotic animals including pandas, rhinos and gibbons were buried with ancient Chinese nobility;[65] and up to 70 million animals (from crocodiles to now-extinct species of shrew) were mummified in ancient Egypt, with around 19 tonnes of mummified cats exported from Egypt to Britain for use as fertiliser (and even as ballast on ships) during the late nineteenth century.[66] In 2017, a porpoise was found in a medieval grave on Guernsey – possibly just to preserve it for consumption, but maybe for more ritualistic reasons. Even more weirdly, the bones of a walrus were found mixed with dissected human remains within a nineteenth-century coffin in the graveyard of St Pancras Old Church during excavation work to expand the Eurostar terminal in 2003. And sperm whale teeth from hunted or stranded animals represent valuable cultural items across the Austronesian world. Whale ivory was traded, carved, and used to make ornaments and decorate important artefacts – and formed high-status grave goods, from ancient Māori burials on New Zealand to being placed under traditional gods' houses on Fiji.[67]

'Do people ever bury animals in graves on Sumba?' I asked Umbu as we headed back to the hotel. 'Yes!' he answered. 'In Mangili, I heard that some people once made a grave for a smart buffalo that went to another village to steal food when their supplies had run out.' And Ron Adams, an anthropologist who did his doctoral research on Sumba's megalithic tombs, reported that snakes are revered in parts of the island and sometimes buried in small tombs. He also documented another tomb containing the remains of a crocodile that was supposedly the reincarnation of a local woman who had drowned.[68] So, this seemed yet another possible explanation for stories of mili monggas and their graves. But the whole thing was getting out of hand. Trying to grasp what might have inspired these tales kept getting more convoluted and complex; it was like chopping off one of the heads of the hydra only for another two to spring up in its place. And this didn't really seem to be helping us find more fossils either.

We decided on one final push to look for evidence of Sumba's prehistoric past. So far we'd focused on exploring the island's eastern fringes and rugged interior; now we headed west. A day's drive from Waingapu brought us to the small town of Waikabubak, where we unpacked our dirty field clothes once more and settled into a hotel off a dark side-street. Again we started the routine of scouting for new caves each day. And there were lots of caves – the cave under the prison, a low flooded tunnel where local women went to wash their clothes; the giant underwater cave behind the rice paddies at the edge of town; and the other caves around Waikabubak, which were also all unhelpfully full of water. The hotel was sandwiched between a mosque and a church; as the sun set each day, the church would blast out funky Christian rock at the same time as the call to prayer. And every evening we had to dodge the gaunt, withered old man who crept silently up the stairs and whispered 'Massage?'

Umbu seemed wary of exploring the nearby hills, making a joke about headhunters when we pressed him. At least, I assumed he was joking. Headhunting had once been widespread across Sumba; it was stamped out by the Dutch authorities (with difficulty) in the early twentieth century, but was an important part of the island's cultural heritage, helping to define patterns of shared ritual allegiance between communities.[69] Although driven by vendettas and local powerplay, ideologically it aimed to capture health and fertility for the whole community, and transform death into life. A *katoda* or skull tree – a wooden altar adorned with heads, usually those of slaves – used to stand in the centre of many villages, the flesh removed from the skulls by cutting, boiling or brief burial. The *katoda* was 'a chilling sight', in the words of a Protestant missionary in 1892; it would supposedly boil with heat and the circle of stones surrounding it would steam with anger when a death went unavenged.[70] There were no longer headhunters on Sumba; but there had been different cultural rules and reasons for headhunting in the east and west of the island, and the two regions were so distinct they didn't even speak the same language. Maybe people from Waingapu were out of their comfort zone here.

Either way, apparently the local hills were a no-go area, so we headed out for a final look for caves in a different direction out of town. We arrived at the edge of a village and followed Umbu along a trail past areca palms and scrubby green regrowth. The trail led down into a narrow gully with rock walls rising up on either side. As we squeezed through a crack under a grassy overhang, the walls opened up around us into a beautiful underground chamber, a peaceful subterranean cathedral lit naturally by a dusty shaft of sunlight that stretched down from an opening in the roof. And the floor of the cave looked like it was covered in sediment – perfect conditions for finding fossil bones.

We spread out and crouched on our hands and knees, scraping away with our trowels at patches of the cave floor. As I dug down into the sediment, though, something didn't feel quite right. 'What's that smell?' Jen asked. The more we dug, the stronger it got. Then I saw something poking out of the ground by my hand – a syringe. I jumped up and looked around more closely. The ground was littered with syringes and other medical waste, and it wasn't natural cave soil; it was human excrement. I yelled out to everyone to stand up and stop poking around in the mounds of shit and needles, and turned to the deceptively beautiful sunbeam; the pile of faeces was directly underneath the hole in the roof. This must be where the locals came to dump their medical and sanitary waste, I realised – unless we'd accidentally stumbled upon the place where the princess from the fairy tale had been brought by her eunuchs. There was certainly no gold belt down here, though. I looked at the excrement smeared down my legs. It was time to go.

Umbu knew of one more cave close to the road on the drive back east. We were all exhausted, but it seemed churlish not to check it out. We parked and followed a trail into the quiet dry forest that stretched back from the road. The flower cones of huge arums, taller than a person and studded with bright red berries, towered over us. A brilliant white paradise-flycatcher darted down the trail ahead, luring me forward like Alice following the White Rabbit. Its magnificent tail feathers, a splay of delicate white ribbons, streamed out behind its tiny body.

The trail dropped down steeply, and suddenly the silent trees opened out and we found ourselves in a vast natural amphitheatre ringed all around with steep cliffs. A small river flowed past, sparkling in the bright sunlight. We were standing in a huge collapsed sinkhole, hundreds of metres across and open to the blue sky overhead. It was secluded and timeless; an unbelievably beautiful lost world. Umbu pointed to an opening in the wall of rock ahead. 'That is Liang Kanabu Wulang. Its name means Falling Moon Cave.' We picked our way across the stones of the river and climbed up to the cave. It was a vast dark cavern; there was no sediment inside, just a jumble of giant muddy boulders and tree-trunks, evidence of periodic flash-floods that must pour catastrophically through the walls of rock and fill the sinkhole during huge storms. We were not alone in the cave; overhead, great leathery black fruit bats clung to the ceiling, mumbling and murmuring to each other as they watched us clamber around on the rocks. We left the cave and sat in the sun, and chewed betel nut in happy shared silence as we looked out over our own private little world. Sumba was an incredible, enchanted place; right now I would have believed any story I was told.

But our time here had finally come to an end. Back in Waingapu I booked some local flights on my laptop, to a backdrop of screams as somebody down the road slaughtered a pig. Tim and James bought some ikat and carved wooden combs as souvenirs, and then we all had a few days of holiday in Flores and Komodo. In Bali, an Australian tourist in front of us in the airport queue was carted off for trying to get through security with a meat cleaver in his backpack; and then we swam with manta rays and turtles, saw Komodo dragons, and watched a Bryde's whale surface right in front of our tour boat as a pair of dugongs dived underneath us in the tropical blue sea. That evening, we sat together on deck as the sun set behind an erupting volcano on the horizon, while thousands of fruit bats flew silhouetted above us through the darkening sky. What an incredible trip it had been; but so many questions were still unanswered – about fossils and endemic species, and the possibility of undiscovered hominins still waiting for us out there in a cave somewhere ... and, of course, all that stuff about the mili mongga. I wondered when I'd be back.

CHAPTER EIGHT

AN INTERLUDE WITH GIANT RATS

As long as we have anything more to do, we have done nothing.

Herman Melville, letter to Nathaniel Hawthorne

Time passed; other work commitments, desperately urgent as always, filled the available hours. Slowly we made progress with sorting, cleaning, measuring and analysing our huge haul of bones. Our initial impressions about the diversity of rodent fossils from Liang Lawuala were correct. Comparison with old taxonomic monographs about rats from other parts of Southeast Asia, Australia and New Guinea, and several days spent examining rodent skulls in the wonderful endless drawers of the Natural History Museum collections, confirmed that nothing quite like our fossils had ever been described. They were new to science. They showed similarities to the unusual giant rats from Flores in characteristics of their teeth and jaws, suggesting that they were most closely related to these animals. But the Sumbanese rodents were not just new species; they exhibited more fundamental morphological differences to rats from other islands, indicating longer isolation, adaptation and evolutionary distinctiveness.

We had two new mammal genera (the taxonomic rank that includes a set of closely related species) to describe and name. The largest new rat, the animal with highly distinctive large fluted teeth, was given the scientific name *Raksasamys tikusbesar* – a slightly cheeky name that literally means 'giant mouse, big rat' in a combination of Indonesian and scientific Latin. Hopefully this name was less of a mouthful than the animal's amazing teeth might have been. The other rat bones represented a second new genus, which we named *Milimonggamys* – 'after the mili mongga, an apparently legendary creature from Sumbanese folklore that was regularly talked about by local people in East Sumba Regency ... combined with "*mys*", the standard suffix for mouse.'[1] At least this allowed us to commemorate the constant presence of the mili mongga throughout our time in Sumba, even if I still had no idea what those stories had been about.

Although different species possess unique adaptations and variation, related species with similar body plans tend to scale in similar ways when they change in size. This makes it possible to infer important biological parameters such as body mass from measurements of specific parts of a skeleton, a mathematical approach that permits reconstruction of general ecological patterns in extinct species and fossil communities. Such scaling relationships have been established for rodents, a group for which many living species exist from which to determine baselines. Using the known relationship between the length of a rodent's toothrow and its overall body mass across a large sample of living species, we calculated that *Milimonggamys* weighed around 300g – bigger than your average black rat, but not by too much (you wouldn't want to wake up with one crawling around in your hair, though). *Raksasamys* was more than twice as heavy, with a maximum weight of almost 700g – comparable to a skunk, hedgehog or small armadillo.

So the giant rats of Sumba were big … but rats from some other islands in Wallacea were even bigger. The surviving giant rat of Flores, *Papagomys armandvillei*, weighs well over 1kg, and some individuals reach 2.5kg.[2] Nage hunters on Flores say the largest rats they catch have a girth greater than a man's calf.[3] Using rodent toothrow–mass relationships we can estimate that the extinct giant rat *Coryphomys* from Timor was also about two kilos. These awesome beasts were similar in size to a small fox. But what's the reason for the striking discrepancy in body size shown by these rats? It's due to the differing sizes of different Wallacean islands. As we've seen, there is a close relationship between an island's total land area – a proxy for its environmental resources – and the corresponding body size of the largest animals it can support. Sumba is a relatively large island, but has a land area of only about 11,000km². In contrast, Flores has an area of over 13,500km² today, and during geologically recent periods of reduced sea level it was connected to several small nearby islands that are currently separated by shallow sea channels, thus substantially increasing its total area. Timor is even larger, covering more than

30,000km². Similar patterns of size variation relative to island area are also seen in other giant rats, such as those of the Lesser Antillean island chain in the Caribbean, which was formerly home to an evolutionary radiation of rodents called rice rats. The largest species lived on Martinique and Grenada, islands with large areas today or during recent Ice Age periods of lower sea level, whereas the smaller islands of St Lucia and Barbados had correspondingly smaller species.[4] In keeping with the typical sad fate of most unusual island species, all these Caribbean rice rats are now extinct.

These examples illustrate the tragedy of biodiversity loss from a scientific perspective. Most of the world's remarkable island species are now gone, and in their absence it can be difficult or impossible to reconstruct the basic evolutionary and ecological processes that regulate how the natural world came to be the way it is. This so-called 'extinction filter' means that any understanding we think we have about patterns of biodiversity on islands based upon what we can still see today – for instance, evolutionary dynamics in island populations – is likely to be both incomplete and biased. This is a basic tragedy of knowledge. By destroying the natural world we can know less about it, and therefore about how to maintain it; and as we are also a product and a part of the natural world, thus ultimately we will learn less about ourselves.

Not all the giant island rats have disappeared, though. *Papagomys armandvillei* is still around, and is considered only 'near threatened' by the International Union for Conservation of Nature (although this assessment is recognised as being out of date). Why has there been such variation in survival between giant rats from different islands, and even between the different giant rats that formerly co-occurred on Flores? We don't know. Maybe island landscapes offered differing refuges from human persecution or the ravages of introduced species, or maybe human pressures differed from island to island due to variation in local cultures or the ways in which people interacted with nature. Or maybe this difference in survivorship instead reflects something intrinsic about the biology and ecology of different species. Were some rats naturally more aggressive, or more likely to spend time in the treetops out of easy reach of hunters? Did they vary in whether they were ecological generalists or specialists, and could characteristics such as variation in their diet have influenced their vulnerability? Or did certain body sizes make some species more or less resistant to hunters or invasive predators? Having a clearer understanding about the cause of this varying resilience would help us predict which other species today might be at high risk of extinction, and thus use evidence to guide appropriate conservation actions. There are numerous fertile research avenues that could be explored to start picking away at these questions, such as studying the ratios of carbon and oxygen isotopes in the teeth of living and extinct Indonesian rodents to reconstruct their dietary niches. But for now, we just don't understand.

The surviving giant rat of Flores is certainly a ferociously aggressive beast, which might provide a clue to how it's avoided extinction so far. Giant rats have a lot of meat on them and are still hunted today by ethnic groups such as the Nage. Sometimes traps are used, but the most effective method is to go hunting on a bright moonlit night when the rats are foraging in the treetops, and use a long bamboo pole to knock them onto the ground where they can then be speared or killed by dogs. But the hunt is very dangerous – 'the angered rodent might land on top of the man or a companion and, with its strong, sharp teeth, tear into his head ... Thus the hunter is advised to lower his head to avoid possible injury to the face or throat'. Giant rats can easily kill hunting dogs by ripping out their throats and can also cause serious injury to people, 'and in common opinion they are the most dangerous of a hunter's quarry'. The Nage consider *Papagomys* to be symbolically linked to the water buffalo as its spirit counterpart because of its size and ferocious behaviour. There is even a specific Nage word, *ji'o*, to describe the giant rat's characteristic method of attacking with its mouth wide open to savage an opponent. Giant rats also appear in several Nage metaphors, of which my favourite is 'to be like a Giant rat's belly' – to 'allude to someone "who receives valuable things but gives only negative things, or then behaves badly," since, as Nage point out, Giant rats eat a great deal but produce only faeces in return.'[5]

Other surviving giant rat species are also respected as an extremely dangerous quarry elsewhere across Wallacea and New Guinea. In *Throwim Way Leg*, Tim Flannery describes a grisly outcome of an encounter with a black-tailed giant rat (*Uromys anak*) witnessed during fieldwork in Papua New Guinea:

> One day on the Sol River a Telefol hunter had come into camp, his hand tightly bandaged. He opened his bilum, and angrily threw down the body of an enormous black rat. Clearly, before examining the catch, I had to treat his wound. As I began to unwind the bandage, I realised that the inner layers were dripping with blood. Its source was a horrific injury to his right thumb. The last joint had been bitten right through, and the nail was shattered with punctures. So severe were these that the end of the thumb appeared to be pulp, which wobbled as I dropped antiseptic onto it.
>
> 'Quotal,' he said, as he explained how he had been feeling in a tree-hollow for a possum. Instead he had come across Quotal, as Telefol know the species. The bite of no other animal is feared as much by them. Its incisors are razor sharp and up to two centimetres long. The terrible damage done by repeated bites to this hunter had been inflicted by an immature rat.[6]

And it's not just giant rats that are surprisingly dangerous – other large rodents can be intimidating when cornered too. In the sixteenth century, Humphrey Llwyd wrote about some of the last surviving beavers in Wales, saying the animal 'hath very sharpe teethe and bytethe cruelly till he perseave the bones cracke'.[7] Indeed, in 2013 a beaver killed a fisherman in Belarus by severing an artery in his leg with its powerful bite, causing him to bleed to death (the unfortunate man had apparently grabbed the beaver and was trying to pose for a photo with it when the attack occurred).[8] More recently, my good friend Ros Kennerley did her PhD research in Hispaniola on threatened hutias (the hefty and deceptively placid-looking rodents only found on islands in the Caribbean). Hispaniola has one surviving species, which weighs over 1kg. One of the people from the village where she was staying wasn't careful when trying to catch a hutia, and the startled animal bit straight through his finger.

Working out why species have died out is much easier if we can understand when they disappeared. What past events were they able to withstand, and can we identify a 'smoking gun' that correlates in time with their extinction? Fortunately, in many cases we at least have the potential to reconstruct extinction chronologies for species that disappeared during recent prehistory. Although most fossils are fully mineralised, animal remains from the past few tens of millennia often still retain some of their original organic biomolecules – notably collagen, a structural protein molecule that comprises over a quarter of the content of fresh bone. Such 'subfossils' usually just resemble very old bones that are still loose within the sediment, in contrast to the fossils you'd normally excavate from rocks with a hammer and chisel. The organic content of subfossils, which provided confirmation of their animal origins, was recognised even during the lengthy historical debates over what fossils actually were. For instance, in 1703 the physician Johann Kerl burned some fragments of mammoth bone and found they produced 'an evanescent urinous spirit along with stinking oil',[9] satisfying Kerl that the remains were organic.

The preservation of collagen within subfossil bones provides the carbon needed for carbon dating. This dating method is based upon the radioactive decay of tiny amounts of a naturally occurring unstable isotope of carbon, carbon-14. This isotope (as a component of carbon dioxide) is incorporated into food chains via plant photosynthesis and consumed by animals during life, and then breaks down after death at a predictable rate over time. The practical upper limit of carbon dating is about 50,000 years, by which point nearly all the carbon-14 within a sample has decayed into stable nitrogen-14. Fortuitously, 50,000 years ago is also about the time that anatomically modern humans first spread out of Africa, meaning that in theory this method can be used to date all past human-caused extinctions.

In practice, however, dating old bones from tropical environments is challenging, because bone collagen degrades and leaches away rapidly under warm, humid conditions, even when buffered through burial within sheltered cave sediments. Even bones that are only a few thousand years old from tropical caves might thus have already leached away all their collagen; they are said to have a high thermal age, in contrast to their actual chronological age. This widespread problem is one of the main reasons why the recent prehistory of many tropical environments is still very poorly understood.

We were in luck though: three specimens that were sent to the radiocarbon lab retained enough organic carbon to permit successful dating. A huge rodent shoulder blade, almost certainly from the giant *Raksasamys*, came back as between 3,950 and 3,700 years old. The two other dates were somewhat younger – a *Milimonggamys* mandible was between 1,950 and 1,800 years old, and a piece of charcoal mixed in with the bones was between 2,300 and 2,150 years old. These varying dates indicated the Liang Lawuala bone bed wasn't deposited in a single rapid event – by one particularly hungry owl – but instead accumulated gradually over a couple of millennia. The mouth of the cave was probably a perfect roost site for many generations of owls, which all spat the undigested bones of their prey down through the cracks in the floor. Studies elsewhere have shown that favoured caves or rock ledges can be occupied continuously by raptors for very long periods of time, providing a fine-scale record of past prey species and faunal change across centuries or more. For instance, analysis of ancient guano deposits reveals that some gyrfalcon nesting ledges in Greenland have been used continuously for 2,500 years.[10]

These carbon dates also provide a last known occurrence date for the giant rats, constraining the period when they must have become extinct. Direct dating of their bones showed that these beasts had survived well past the first arrival of modern humans in Southeast Asia, and beyond the disappearance of the hobbit on Flores and the different tiny stegodons across Wallacea; they had made it through the devastating switch from glacial conditions to modern tropical environments at the end of the last Ice Age climatic shift around 11,700 years ago; and they had persisted alongside prehistoric human colonists on Sumba through most of the current Holocene Epoch. There were still giant rats on Sumba when the Roman Empire spread across Europe, when people in China started using paper and the wheelbarrow, and when the Kama Sutra was written in India. Based on our sample of just two dated subfossils, it's extremely unlikely that we somehow managed to find and analyse the bones of the very last surviving giant rats; as we've seen, this assumption would be a Romeo Error. Our limited series of carbon dates will thus almost certainly pre-date the actual extinction of these species. But we could now say that giant rats were definitely still living on Sumba into the last few millennia – tantalisingly close to us in time. So what might have caused these remarkable animals to disappear?

The recent fossil record reveals that humans have been associated with biodiversity loss for tens of thousands of years, ever since the Late Pleistocene Epoch. The world was then a very different place. It was still in the cold, dry grip of the most recent Ice Age glaciation – and all but the most inhospitable polar and desert landscapes were home to diverse communities of huge terrestrial animals. These large-bodied mammals, giant birds and reptiles were collectively known as megafauna. They ranged from familiar prehistoric Ice Age beasts such as mammoths, mastodons, woolly rhinos and sabre-toothed cats, to South American giant armoured glyptodont armadillos that weighed 1.5 tonnes, and bizarre Australian marsupials such as palorchestids, diprotodonts, pouched 'lions', and giant short-faced kangaroos almost 3m tall. This abundance of megafauna was the evolutionary norm; despite constant change in environmental conditions over geological time, global ecosystems had supported a diverse mammalian megafauna for tens of millions of years, and before that the world's terrestrial landscapes had been dominated by dinosaurs for an even longer period.

A handful of megafaunal species are still just hanging on today – three species of elephants, five species of rhinoceroses, and a few other giant terrestrial mammals such as wild Asian cattle – but these are the few survivors of what's been called an 'eco-catastrophe'. As modern humans spread out around the world and reached each of the major continental landmasses, the megafauna vanished in their wake. In total about 100 genera of megafauna became extinct during the Late Pleistocene. This was also when all the other species of humans found in different parts of the world – the Neanderthals, the Denisovans, and the tiny island people of Flores and Luzon – seem to have vanished too; in order to understand their extinction, it might be instructive to view these species as other examples of mammalian megafauna as well. In the poetic words of Alfred Russel Wallace, 'We live in a zoologically impoverished world, from which all the hugest, and fiercest, and strangest forms have recently disappeared' (although he went on to add that 'it is, no doubt, a much better world for us now they have gone', a sentiment that most conservationists would take issue with).

What actually happened has been the subject of intense – and sometimes acrimonious – debate among palaeontologists and archaeologists since the nineteenth century, because the story is complicated by a second 'smoking gun'. These prehistoric human migrations and megafaunal extinctions all occurred during an interval of severe natural climate change, which has confounded efforts to determine causality. The end of the Late Pleistocene saw the Ice Age cycle shift rapidly between different equilibrium states, from glacial conditions to more temperate environments. Global temperatures increased extremely fast, with shifts reconstructed from Greenland ice core samples of up to 10°C over a few centuries or even decades.[11] This abrupt and violent increase was followed by further large-scale destabilising wobbles and reversals before broadly 'modern' Holocene conditions were reached. Ecosystems around the world thus experienced sudden stresses and shifts to which slowly-reproducing large animals would have been intrinsically vulnerable, because their populations required large amounts of stable resources and could not respond quickly to fast-changing conditions. Wallace himself swung back and forth over whether he thought megafaunal extinctions were caused by climate change or prehistoric humans, and opposing entrenched camps of academics have marshalled varying lines of evidence for both hypotheses ever since.

The first champions of the human-caused extinction model advocated the 'overkill hypothesis', whereby megafauna were wiped out rapidly via overhunting by small bands of prehistoric hunters following first contact. Megafauna need not have been preferentially targeted, or even hunted that frequently; population modelling shows that, thanks to the long lifespans and low reproductive output of large animals, just knocking out a few here and there (especially young individuals, which might have been easier to hunt) would have been sufficient to make them vanish within a very few centuries.[12] Subsequent arguments have suggested that extinctions were possibly more gradual and caused by indirect as well as direct human impacts, such as ecosystem modification through large-scale burning of vegetation. Increasing availability of good-quality dates, obtained through carbon dating and other methods, has shown that the last occurrence of megafaunal species across different parts of the world typically correlates with the spatially staggered pattern of prehistoric human arrival in different regions – starting with Australia by 45,000 years ago and not reaching the Americas until tens of thousands of years later – instead of with the known timing of climatic switches. Ultimately, the whole 'either/or' debate around causality of these extinctions is probably something of a false dichotomy. Although megafauna had persisted through many previous Ice Age shifts, suggesting that ancient climate change alone was probably insufficient to drive extinctions, their populations would have been naturally stressed at this point – and thus, unfortunately, probably easier to wipe out by newly arrived humans. It would have been a fatal synergy.

The disappearance of most of the world's megafauna during the Late Pleistocene set the template for what was to come. It is sometimes suggested that there was a lull in human-caused extinctions for much of the Holocene, before modern-day destruction of the world's ecosystems really kicked off. This idea of a 'ceasefire' period was termed 'Holocene underkill' by archaeologist Donald Grayson, to contrast with the concept of Pleistocene overkill.[13] However, the idea of an idyllic, Eden-like time when humans and wildlife coexisted in harmony doesn't match evidence from the recent fossil and archaeological records. Although most of the megafauna were already gone by now – meaning, bluntly, that there was less big stuff left to wipe out – considerable further biodiversity loss occurred throughout the Holocene. Whereas climatic and environmental conditions have been relatively stable for the past 11,700 years (at least in comparison to the massive climatic spasms at the end of the Late Pleistocene), human cultures around the world changed from small-scale, low-density hunter-gatherer communities into larger settled agricultural communities, exerting progressively more pressure on local landscapes and biodiversity. Eventually, many agricultural communities developed into urbanised and industrialised nation-states, accompanied by exponential population increases and even greater demands for natural resources. More than a thousand years ago, cities such as Rome, Baghdad, and Chang'an in central China already had over a million inhabitants, and forests around the world from Asia to the Andes were being cut down and burned to facilitate agricultural expansion long before recent history.

Thanks to this constant long-term modification, most of the world's landscapes are now 'non-natural' to some degree. Even seemingly pristine wildernesses such as Amazonia are recognised as 'built landscapes' that were profoundly shaped by human actions throughout recent millennia, such as enhancement of soil nutrients to support food production, and alteration of plant distributions and densities to encourage the growth of useful species such as Brazil nut trees.[14] American palaeoclimatologist William Ruddiman champions the hypothesis that early agriculture even caused an increase in greenhouse gases, which had global effects on temperature during prehistory.[15] In particular, rice paddies emit the greenhouse gas methane after rice is harvested; and levels of atmospheric methane are known to have risen unexpectedly around 5,000 years ago, just at the time when rice agriculture spread across eastern Asia. Comparable increases are shown by atmospheric carbon dioxide in the mid Holocene, again potentially associated with changing land use by prehistoric human societies. Ancient human activities might therefore have initiated unnatural climate change much earlier than we've generally thought.

Unsurprisingly, this escalating increase in human pressures was associated with many further species losses. The north African long-horned buffalo and Chinese short-horned buffalo; Europe's wild ass and giant deer; unique ducks and turkeys in North America and herons in the Middle East – all these species, and many more, survived into the Holocene but died out before recent history. About 200 different mammal species alone are known to have disappeared during this interval, with many extinctions occurring on islands that were first reached by human colonists during the Holocene.[16] Other species that avoided complete extinction vanished from large parts of their former Holocene ranges. Lynx, elk, pygmy cormorants and Dalmatian pelicans were present in Britain until a few thousand years ago, wildcats once ranged as far north as Sweden, and Asian elephants could be found from Syria to northern China. As there have been no major climatic shifts during the Holocene, it's safe to assume that these species losses and population collapses were all caused by some sort of human actions during recent prehistory.

Even Indigenous cultural groups, perceived as being in tune with their environment today through their traditional practices, have often undergone considerable changes in the ways they have interacted with local biodiversity over time. For example, multiple lines of evidence indicate that Aboriginal societies across many parts of mainland Australia intensified their use of natural resources around 4,000 years ago; more sophisticated stone tools such as bifacial projectile spear points, spear barbs, and millstones for grinding seeds appear in the archaeological record, and people started to consume a wider variety of plants, began continuous inhabitation of previously seasonal sites, and started to occupy deserts, alpine regions, rainforests and mallee woodland for the first time.[17] The dingo was also introduced to Australia around this time, indicating some sort of trade or other interaction with sea-going people from Southeast Asia. The timing of Australia's regional cultural intensification matches the known timing of the demographic expansion of Austronesian-speaking farmers across the nearby islands of Wallacea and the tropical Pacific, raising the possibility that interactions with these sea-going colonists triggered a cultural shift in Aboriginal societies even if Austronesians didn't settle on the continent themselves. This striking suite of cultural changes is associated with the disappearance from mainland Australia of thylacines, Tasmanian devils, and a plump flightless bird called the native hen. All three species survived into recent times only on the isolated island of Tasmania, where local Aboriginal cultures did not undergo a comparable intensification of natural resource use.

This look back across the history of human interactions with biodiversity provides new perspectives about what might have caused the disappearance of the giant rats of Sumba. Could prehistoric human hunting or habitat loss, which have been driving species to extinction for millennia, also have wiped out these native island rodents? People on Sumba today still hunt and eat the island's surviving mammals, as is the case in rural communities in tropical countries around the world. Various methods are used: local hunters living near Liang Lawuala had visited nearby caves to knock fruit bats down from the ceiling, and in her book *Ignition Stories*, anthropologist Cynthia Fowler describes an encounter with Sumbanese hunters near Kodi who were smoking fruit bats out of palm trees, and were clearly getting excited at the prospect of eating them:

"Hi, what are you fellows up to?" asked Dita Katodo.
"Bats live in these trees. So, we're burning this palm tree to smoke out the bats," one of the men answered.
"Mmm. Bats," said Bilya Peghe.
Bilya Peghe's granddaughter said, "Oooh. Tasty."
"Been a long time since I've eaten bats," said Raddu Palla.[18]

Sumba has the lowest human population density of any of the main Indonesian islands of East Nusa Tenggara, and has also experienced the lowest population growth since the mid nineteenth century – a rate of only 0.01 per cent. The reasons behind this lack of growth are unclear, although anthropologists have speculated it might be linked to the very high ceremonial costs associated with weddings in Sumbanese culture.[19] Given the island's human density, would direct exploitation of native species by local people therefore really have made much of a dent in animal populations?

The use of fire to hunt bats on Sumba points to another possible extinction driver for the giant rats, though. During our long journeys back and forth across the island to look for caves, the landscapes we drove through were dry badlands and treeless savannas. Some of this modern-day expanse of grassland must be a natural part of Sumba's pre-human environment. The island is geographically situated in a natural rain-shadow so receives little precipitation; some of its savannas have botanically complex herbaceous layers, suggestive of a long evolutionary history; and its avifauna includes an endemic species of buttonquail, a shy and enigmatic little bird that spends its time skulking in dry grassland habitats (and which I had frustratingly failed to see, despite trudging through a lot of grass looking for it). However, savannas are also what forested ecosystems turn into if they're burned – an easy method to clear trees, especially in dry environments. Periodic burning will then maintain these open landscapes for agriculture and provide fresh grazing for horses and cattle, and free-roaming livestock will also nibble away any regenerating tree seedlings that might have survived in the soil's seed bank.

Fire management practices – so-called 'ignition ecology' or 'firestick farming' – are employed widely across Wallacea and by other Indigenous communities around the world. Landscape burning has even shaped British landscapes since long before recorded history, with our heather moorlands and acid grasslands both generated and maintained by regular burning rather than constituting 'natural' habitats. The potential ecological consequences of this landscape management approach vary from one ecosystem to another. Although burning stimulates new growth of non-woody plant tissue, and is thus associated with an increase in above-ground nutrient availability and productivity (which is why it's a good temporary fix for agriculture), anthropogenic grasslands across Wallacea support relatively little biodiversity compared to the tropical forests they have replaced, and their exposed landscapes usually experience massive soil erosion – a critical problem for tropical soils, which are often naturally thin. Even if left alone, it's estimated that these drastically damaged systems might take thousands of years to recover naturally, and they have been described as 'green desert'.[20] Conversely, periodic rotational burning by Aboriginal communities in Australia created ecosystem mosaics of open and regenerating vegetation, which appear to have mimicked the natural landscape gaps once generated by the continent's now-extinct megafaunal grazers and browsers. Indeed, the restriction of Aboriginal firestick farming by Europeans is now regarded as a key factor responsible for the historical extinction of many of Australia's endemic smaller mammals, including the crescent nailtail wallaby, lesser bilby and desert bandicoot, which appear to have been naturally dependent upon such mosaic habitats.[21] However, landscape burning is also suggested as the potential mechanism by which the first human settlers eliminated Australia's megafauna 45,000 years ago.

The widespread practice of burning landscapes in the Lesser Sundas was recorded as early as 1699, when English buccaneer and explorer William Dampier observed that the inhabitants of Timor 'take but little pains to clear the land; for in the dry time they set fire to the withered grass and shrubs, and that burns them out a plantation for the next wet season.'[22] Almost a century later, Captain Cook and Joseph Banks passed close by Timor during the HMS *Endeavour* voyage, and reported that 'many fires were also seen on all parts of the hills, some very high up', even when they were 'too far off to see any thing but large fires of which were several ashore.'[23] Visitors to Sumba in the early twentieth century recorded seeing cut tropical forest tree stumps, tree-ferns and burnt tree trunks left as isolated relics within newly-created grasslands,[24] and the island's surviving forests are now almost entirely restricted to high elevations, or to ravines and gullies that are too steep to access or plant crops. Landscape management by firing grasslands is a ubiquitous practice across Sumba today, and although the island's environmental history is poorly understood (hence the motivation for our fossil-hunting visit), fire is likely to have been used locally for millennia to clear forests and maintain open habitats. It's hard to reconstruct Sumba's original pre-human forest cover, but the island has probably been shaped by ongoing anthropogenic disturbance for so long that its remaining forests might now constitute just a few per cent of their past distribution. This long-term legacy of environmental destruction was further compounded by historical demand for slow-growing sandalwood trees, which led to further widespread clearance of forests across Sumba from the 1700s onward. Nearly the entirety of this remote island is thus now an artificially maintained anthropogenic ecosystem, rather than anything close to its pre-human state.

Although they're often considered separately, hunting and habitat destruction are not mutually exclusive threats. Opening up landscapes also allows increased access for hunters, and it can be difficult to unpick the relative contribution of each process to a species' extinction event. Firestick farming in Australia was largely conducted by Aboriginal communities to enable more efficient foraging of small game species adapted to open landscapes, such as goannas; indeed, it can be considered as a method of 'farming' these species within deliberately constructed 'pyrosphere' landscapes by maintaining their populations at high densities. In addition to clearing native forests for agriculture, fire is also used to facilitate hunting in parts of Sumba today, such as in the region around Kodi.[25] Hunts are organised every year to catch the invasive rats that now occur across the island's human-modified ecosystems; they typically take place at the height of the dry season in August or September, when the savanna grass burns most fiercely. As in Australia, rotational burning is practised, with different patches burned each year to allow the grassland to recover.

Villagers make torches from bundles of dried grass and wood, and start fires along the periphery of the patch of savanna where the hunt is going to take place. Coconut fronds are used to manage the course of the burn and stop it from spreading uncontrollably, and rats are caught as they try to escape the flames, or are trapped and dug out from burrows that are exposed when the tall grasses and shrubs are burned away. The captured rats then have their hair burned off and are gutted and roasted. In contrast to the giant rats on Flores, which are said to taste like water buffalo,[26,*] rats on Sumba reportedly taste 'like dog meat'.[27] This local custom provides an intriguing possible scenario for how people on Sumba might also have once interacted with the island's native rats. Maybe the very last *tikus besar* met its end after rushing from the flames of a deliberately started savanna fire into the hands of a waiting hunter.

It's possible that Sumba's giant rats succumbed to gradual attrition; maybe ongoing human encroachment over many millennia eventually crossed a threshold of sustainability. It's also possible that some novel destructive force arrived on the island during recent prehistory, giving a shock to the system that suddenly tipped the balance from survival to extinction. The different waves of prehistoric human colonists that reached Sumba at different times are likely to have interacted with their environment in different ways. The first hunter-gatherer settlers probably existed at very low population densities, much lower even than seen today on this sparsely inhabited island, so might have had a relatively low impact on landscapes across Sumba. However, our carbon dates for *Raksasamys* and *Milimonggamys* are close to the time of the 'Chinese steamroller', when Austronesian colonists spread across Southeast Asia. The arrival of these Neolithic farmers on Sumba must have had dramatic environmental consequences; wide-scale deforestation, to clear land for crops needed by larger agricultural human populations, might only have started in earnest once they arrived. And all the rodent bones at Liang Lawuala were from endemic rats – the owl that made the bone deposit hadn't eaten any invasive

---

* The reported flavour of Flores giant rats may in fact be associated with the wider symbolic linkage between giant rats and water buffalos within Nage cosmology.

rodents, presumably because they hadn't got to Sumba yet. The timing of the arrival of invasive rodents across Southeast Asia is poorly understood. Pacific rats might have been introduced to Sumba by Austronesian colonists as part of their travelling prehistoric menagerie that also included exotic pigs, civets and macaques. The arrival of other invasives probably occurred more recently, maybe sometime during the past few centuries as international trade increased across the region, although possibly earlier, as Chinese traders might have been visiting nearby Timor for sandalwood as long ago as the fourteenth century.[28] But irrespective of when they arrived, the invasives almost certainly had a major impact on Sumba's native biodiversity. Did they bring about fatal declines in the giant rats through direct competition, or indirect competition for resources such as food or denning sites? Or did they carry novel pathogens with them, like the trypanosomes that wiped out Christmas Island's native rodents? Any and all of these harmful interactions are possible. Maybe the final end was the result of a synergy of pressures, both quantitative increases in long-term human activities such as hunting or habitat loss, and qualitative changes in the types of threats that were present. Our discoveries and handful of carbon dates were just the start of our understanding about Sumba's past.

As we prepared a scientific paper reporting our discoveries from Liang Lawuala, we learned of a remarkable coincidence – a few years earlier, a team of Australian and Indonesian palaeontologists led by Gert van den Bergh, one of the researchers who worked on the hobbit fauna at Liang Bua, had also visited Sumba to see if they could find any fossils. They hadn't uncovered any cave deposits, but instead had heard about some bones found by a man digging a well in his backyard in Lewa, a small village of a few houses stretched along the road running through the centre of the island. At the bottom of the pit they'd excavated the jaw, part of the forelimb, and a few other fragments of a juvenile stegodon. And alongside the stegodon bones they'd discovered something even more exciting – a large curved tooth, transversely flattened and serrated to enable it to cut effortlessly into flesh. It was the fossilised tooth of a Komodo dragon – the first evidence that Sumba had also once been home to these giant beasts, which now survived only on the western tip of Flores and on Komodo and neighbouring islands to the north. Was this tooth accidentally shed thousands of years ago by a dragon that was feeding on the remains of the dead stegodon?

Based on uranium-thorium dating, a technique that utilises the natural radioactive decay of unstable uranium and thorium isotopes across a longer time interval than the decay rate of carbon-14, Gert's team estimated the bones from the well were over 100,000 years old. These fossils provided even more evidence that Sumba had once supported a very similar fauna to that of Flores. Once upon a time, these two nearby islands both had upside-down ecosystems containing dwarf elephants and giant rodents, which were predated by huge reptiles. And these additional discoveries made it seem even more likely that there had once been some sort of 'hobbit' on Sumba, too. All we needed to do was keep looking.

Gert and his collaborators co-authored a joint scientific paper with us, including descriptions of the Pleistocene bones from Lewa and our Holocene cave fauna from Liang Lawuala, providing an overview of what we now knew about Sumba's prehistoric past.[29] But I still had unfinished business on Sumba; there was so much more to learn. It had been a long time since our visit. I wanted to go back.

The new Sumba team met up in an airport hotel in Jakarta, which had a real live bird of paradise in the lobby and a café that served drinks with names like Brandon's Flaming Avocado Juice. Tim had been able to come back with me, but neither Jen nor James could accompany us on this return trip. Instead, we were joined by Paul, another one of my doctoral students – a laconic and very amiable young man who was conducting anthropological research in Papua. We got through our jetlag in Bogor, a pleasant leafy city situated inland from Jakarta, which had been the old colonial administrative centre and the governor-general's summer residence when Indonesia had been ruled as the Dutch East Indies; it had then been called Buitenzorg, or 'no worries' in Dutch. We sat in a café inside the stunning botanical gardens as a thunderstorm broke overhead. I watched the rain pour through the thatch, and savoured the pungent, earthy smells released from the tropical vegetation around us by the downpour. Paul drank something called kedongdong juice, Tim ate a durian ice cream, and I had a beer. It was time to move on; Sumba was calling.

CHAPTER NINE

THE ISLAND OF THE DAY BEFORE

(or, The Soot Bill and other Half-Remembered Tales)

It is apparent that the modern biologist who ventures an excursion among the peculiar idioms of an ancient people whose scientific zoölogical terminology he does not understand may make the Olympian Jove shake the empyrean with his laughter.

A. H. Godbey, 'The Unicorn in the Old Testament'[1]

Without any warning, the old man reached into the small receptacle beside him and threw a handful of rice into my face. He turned and flung more handfuls into the corners of the hut and then intoned with authority for several minutes, periodically throwing rice over his shoulder. His voice bubbled from his toothless mouth, and the gloomy light of the hut softened his deeply lined face. He handed out some betel nut; I tried to aim the constant flow of red spit that gushed out of my mouth down between the gaps in the bamboo platform on which we were sitting, but it ended up everywhere – all over my trousers, on my notebook and over the floor. I wondered vaguely whether Indonesian washing powder adverts boasted that they could get betel stains out of clothes. I felt dizzy and relaxed, as though I'd mixed alcohol with coffee. Everything felt a bit slowed down. 'And now something to drink!' said Umbu, handing round cups of palm wine.

From the air, the dry yellow landscape of Sumba seemed as ancient and timeless as ever. As we flew past the sparkling blue plain of the Sumba Strait, the hills along the northern coast formed a series of huge flat steps that fell down towards the sea; these were ancient coral reef terraces that had been frozen in stone when the island had slowly risen up from the water a million years ago. There were a few more foreigners arriving with us this time – we stood alongside a group of rowdy Australian surfer dudes and an elderly lady wearing an enormous floppy sun hat as we waited for our luggage in the tiny airport, while a series of cardboard boxes and clingfilm-wrapped Hello Kitty suitcases trundled round on the conveyer. We walked out of the airport into a blast of dry, hot air, and a further blast of yelling taxi drivers grabbing our bags from our hands and throwing them into cars before we realised what was happening. After an intense burst of rapid haggling and shouting (no, each foreigner does not need to go in a separate taxi), we were on our way once more. Tim leaned forward to talk to the driver we'd ended up with. 'Do you know anything about the mili mongga?' he asked. Paul chuckled. 'No, man! What is that – a *giant*??' our driver roared with laughter. 'I'm a Waingapu boy – we don't know about that round here!'

The lady at the front desk of the hotel gave us a huge smile as soon as she saw us. 'We meet again.' Everything in Waingapu was reassuringly familiar. The little shop selling cakes and adult diapers was still there round the corner, and a billboard in front of the local hospital advertised cures for betel nut addiction. An ice cream seller banging a gong cycled past on a rickety bicycle, balancing a huge metal container labelled 'Es Krim' on the handlebars. And a group of peddlers sat amongst the dried stains of betel spit at the side of the dusty road. One of them held out some carved dugong teeth.[*] 'A dugong got caught in a big net near Waingapu,' he said. 'If you can collect the tears of a dugong, even just a single one, you will be lucky every step you take for the whole of your life. So everyone from Waingapu went to see the dugong, and cut it into pieces.'

We had dinner with Umbu and a friend at a tiny eatery on the waterfront. 'Welcome back Pak Sam!' A huge grin broke across Umbu's face as he shook my hand in greeting. We chatted for hours, savouring the warm night breeze coming in off the sea as Umbu talked about what had happened since our previous visit. Indonesian country and western music played softly in the background, and Tim smoked kretek, the local clove cigarettes. 'Tim is so *nakal!*' Umbu teased. 'Naughty': my first new Indonesian word of the trip.

---

[*] In the 1920s, Karel Dammerman reported that people on Sumba hunted dugongs for food and carved cigarette-holders from their tusks.[2]

Umbu had continued asking around about mili monggas. There were more stories, but he still hadn't found anyone who'd seen a mili mongga themselves – all the supposed witnesses had died decades ago. People were moving away from the remote forests now, for better farming opportunities or improved access to water and other resources … so maybe that was why mili monggas were not seen much anymore. We talked about Sumba's other local mysteries, such as the roh kurcaci, a naughty entity resembling a bald gnome that only wore shorts. And Marapu was now in the process of becoming legally recognised as an official religion following judicial review; many people were converting back to it from Christianity. Under Marapu traditions it was forbidden to cut down forests around natural springs, and other areas of land also had to be set aside and left as forest. Not every tree could be used to build a house, and a shaman had to be consulted to examine the entrails of a pig or chicken before deciding which areas of forest could be cleared for farmland. These traditional practices might have the potential for a more sustainable coexistence with what was left of Sumba's natural environments.

We set off once again for West Sumba. Theodorus Lambertus Verhoeven, the Dutch missionary and archaeologist who originally discovered Liang Bua cave in the 1950s, had also visited Sumba in 1956 to look for fossils. In a brief paragraph tagged onto the end of a German-language paper about the archaeology of Timor, Verhoeven had mentioned finding some prehistoric artefacts in West Sumba similar to those from Timor and Flores, including ancient tools made from seashell (*Muschelschalartefakten*) and stone. To aid the interested reader, he provided a map showing the locations of nine caves in which he had found artefacts.[3] However, the tiny hand-drawn map was not quite the break we needed; it showed a few numbered blobs clumped together on the western end of the island, with place-names rendered in an old-fashioned phonetic guesstimate that Umbu couldn't make sense of. Every fictional treasure map I'd seen as a kid, from the map of Maple White Land in *The Lost World* to One-Eyed Willy's map in *The Goonies*, was better than Verhoeven's. This was going to be an adventure.

Unsurprisingly, the map left something to be desired when we tried to use it to navigate. Eventually we ended up in a tiny village somewhere to the west of Waitabula – another row of wooden houses with steep corrugated metal roofs lining a dusty road. We were directed to a slightly larger house to the right of the road; plastic chairs, coffee, betel nut and cigarettes appeared, and I showed my crumpled photocopy of Verhoeven's map to the middle-aged woman who lived there. She started to go through the 'Grottennamen' tongue-twisters as Tim got stuck into the betel nut. '"Lete Kaghona"? That's not a cave around here. "Wihitkaputjah"? Maybe that's We'ekapioto. That's a vertical cave – you drop into it and it's full of water. It's very dangerous.' A very old man emerged from the house behind us to watch the proceedings with indifference.

'I own a bat cave near here with a sacred snake in it,' the woman continued. 'It comes out on a full moon. It's got horns and its body is as thick as a tree. And there's another cave, further away, which contains something that if you try to take a photo of it, it will break your camera. It's something Hebrew or Aramaic. There used to be biblical giants that walked the Earth – that's what the mili mongga is.' She chain-smoked as she expounded her unorthodox views on history and time. 'First of all the Portuguese came to Sumba, about AD 1500, so that's a thousand years ago. Then Genghis Khan arrived from Mongolia and brought other faiths with him. Maybe when Titus destroyed the temples and the Jews spread around the world, they also came to Sumba, because there are Aramaic scrolls at Beindello village. You should tell Israel that they're here. Africa and Sumba have similarities too, because in the Congo there's also a place called Sumba.' She seemed obsessed with Leviticus. Well, it was a Sunday.

One of the villages nearby had a familiar-sounding name: Migurumba. It felt wrong not to go and visit; and inevitably it was harder to find than expected. We drove along miles of featureless dirt track, a Coldplay CD on repeat in the car stereo until we knew all the lyrics whether we wanted to or not. Somewhere on the way we passed a small house with two dead Brahminy kites nailed to the wall.

Migurumba village was stunning. Tall thatch roofs stretched up all around us, rows of blackened pig jaws hanging from their rafters; there wasn't a corrugated metal roof in sight. A group of megalithic tombs nestled together on a green lawn, adorned with colourful paintings of Jesus. Umbu explained that they were made of concrete because the large limestone slabs preferentially used for capstones could only be found in East Sumba. A crowd of people gathered round us. Everyone was smoking. 'This is a sacred village,' Umbu said. 'The Dutch tried to burn it down, but it couldn't be burned because it has sacred protection. And they know all about the mili mongga here – they call it migurumba.'

The rato, the village's spiritual leader or shaman, was tiny, with his grey hair poking over the top of the strip of brightly coloured cloth wound around his head. We climbed into his hut and sat cross-legged on the slats of the raised bamboo platform under rows of pig jaws. A fire was burning somewhere further inside the hut. Most of the village crowded in around us in the darkness; one of the villagers sitting opposite started filming me on his phone. The rato laid out seven wooden plates. 'There are seven plates for the seven families in the village. If you want to learn what the rato knows about the mili mongga, you will have to put 200,000 rupiah, some betel nut and some rice on each plate, so that each of the families receives an offering. In fact, it should be 177,000 rupiah; that's the powerful sacred number of the nyale worms. But the rato has rounded it up for you so it's easier to pay.' I had a quiet discussion with Paul and Tim as the rato and the villagers waited in silence. The price was about the equivalent of a night in an airport hotel in Bali (and not even one of the posh ones where you could re-centre your cosmic aura before hopping on a flight). Wanting to know whether the mystery of the mili mongga could really be cleared up this way was just too tantalising. I dug out my wallet.

Once the rice-throwing was over and we had all knocked back healthy amounts of palm wine, Umbu translated as the rato began. 'The migurumba lived in this village a long time ago, before the Dutch came. He used to hunt and eat humans. He was over two metres tall, and was covered with long blond fur. He didn't wear clothes, just underwear made from tree bark.' The bamboo planks were uncomfortable, people were packed in all around me on the cramped platform, and I felt spaced out; it was hard to think of any questions to ask. 'He was born over where those big trees are. His parents were human, so he was really still a kind of human too, even though he could live for over a hundred years. The parents didn't do anything wrong to deserve having the migurumba as a child; God just sent him to them. He couldn't speak – he could only make a sound like "*kik, kik*'" – the rato paused to impersonate the noise – 'and he couldn't understand normal speech. He was called Migu, but when he grew up and became furry, people added "rumba" onto the end of his name, because it looked like he was covered in grass.' (The Indonesian word for grass is *rumput*, although the word for human hair is the very similar-sounding *rambut*.) The rato scooped up a mixture of betel nut and lime from a little pot on his lap and pushed it into his mouth, scrunching the mixture with his gums. 'The people asked the migurumba to leave, so he went to another village. He killed so many people that everyone lost count. He ate them raw and sliced them into little pieces with his long claws. He ate anyone – he wasn't choosy. The villagers couldn't do anything – they were too afraid. Nobody had spears or parangs or guns in those days. Then in the end, he got old and one day he got sick and died. He was buried outside the village. You can go to see his grave if you like.'

Hmmm. Compared to the more naturalistic stories about mili monggas we'd previously been told in East Sumba, this just felt like a myth or fairy tale about an evil giant who lived long ago, with an anticlimactic ending and a forced explanation for the monster's name. It seemed no more 'real' than the folktales collected by anthropologist Janet Hoskins from this part of Sumba, some of which also featured aggressive hairy giants alongside other magical entities and fantastical happenings.[4] Then the rato reached towards me with something from deeper inside the hut. 'Now you have to sign the guest book!' It was open at a list of comments and signatures. 'Beautiful village – we had a great time!' 'Lovely to get to see some of the local culture on our holiday.' Tourist groups from France and the Netherlands had visited in the past few days. 'And now you need to make a donation to the village.'

I might as well try to get my money's worth, I thought. 'If the migurumba was so bad, why did the village get named after him?' I asked the rato. The old man lit a cigarette.

'The village has another real name,' Umbu explained, 'but it's famous because of the migurumba.'

'But if he was so bad, then why was he buried?' I continued.

'Even though he killed people, he still came from human parents and so it was right to bury him. But if he'd been buried here, the people would have been afraid that lightning might strike the village, so he was buried at the edge of the kampung. There's nothing special about his grave, no magical powers or anything like that. He was buried without a big stone on top, though, because he did eat people. No ceremonies are ever made to him, and the grave has always been left undisturbed – nobody has ever looked inside.'

The rato had been talking for an hour. One of the villagers passed round some damp biscuits and small cups of coffee that had clearly been made with salt water and then flavoured with huge amounts of sugar, making it taste like toffee. I chewed on the damp biscuit. 'When did this all happen?' I asked.

'This house is the oldest in the village,' Umbu translated. 'It was already standing when the migurumba's parents' house was built. And the village dates back to the first time the Pasola was held. The people tried to hold the Pasola in Bukubero village, but the ground was too slippery there. To work out the best place to hold it, a bow and a spear-thrower were made out of a palm branch and a competition was held to see who could hit a target that was a human hair. Seven men, all ratos from four kampungs, competed in the games,' – the rato paused to list the competitors' names – 'and the seventh man won and built this village. Bukubero village gave them a plate of nyale worms and the Pasola was held here instead. The rato is the fourteenth-generation descendant of the winner. The migurumba story happened sometime within the past fourteen generations.'

This date was more vague than the stories we'd heard in East Sumba, which had allegedly taken place only about a century earlier. But the conversation around us became animated as the villagers crouching in the rato's hut began an excited discussion about the archery contest. A lengthy and heated debate commenced over the names of the seven contestants, who seemed to be the same seven families to whom we'd had to make offerings. This was of much greater local importance than a story about a mere cannibal wildman. The rato began to list the names of his family lineage from the winner of the first Pasola onward. All this begatting felt very biblical. 'Not everyone knows how to trace their heritage like this,' the rato explained. 'But it is important to try, for the traditional ceremonies. Some people are even able to list their ancestry all the way back to Adam when they are grieving at an important funeral.'

When the conversation eventually concluded to everyone's satisfaction, we thanked the rato and the villagers and jumped down from the hut. The rato lowered himself down gently after us. The effects of the betel nut, palm wine and coffee had worn off, and it was a bright sunny day. 'What's the best way to get to the migurumba's grave?' I asked. One of the villagers took Umbu's arm and spoke to him quietly. 'Pak Sam, the grave is too far away for us to visit. It's very difficult to get to.' Ah; we'd heard this all before.

We walked back towards the concrete tombs at the edge of the village where the car was parked. Two villagers were bent down beside the nearest hut, washing a contented-looking pig. Someone else started talking to Umbu. 'A man called Rangamoni is buried by the old migurumba house. He was the migurumba's friend; because the migurumba was human, someone needed to be his friend. His grave is unmarked, but this man can show you where it is.' We walked to a patch of rubble and overgrown vegetation. 'You need to stand on his grave and have your photo taken!' shouted Umbu. I posed on the pile of rubble, feeling self-conscious. Some boys from the village stared at me. 'Rangamoni was buried in secret, with gold and a horse statue,' Umbu relayed. 'And the earth and sky were closer together back then, so he climbed up to heaven on a palm tree.'

This was another variation on the relationship between mili monggas and graves that we kept hearing about. But most of these graves turned out to be unmarked, or suddenly inaccessible; how many were even real? And although I wanted to explore new ways of learning about the past to understand environmental change, could a local story about an event that supposedly happened fourteen generations ago – or even a hundred years ago, for that matter – really have any basis in fact? In a culture lacking historical birth or death certificates, or the equivalent of written parish records or other documentation, it didn't seem possible to verify the rato's impressive claims about his pedigree, or to understand whether these antecedents were instead being viewed through the lens of mythology. Interpreting wild stories from Sumba and beyond about mili monggas, meu rumbas and ebu gogos depended upon a crucial assumption – that rural communities were able to preserve accounts of important past events as part of their oral history, through a process of cultural transmission down the generations. But the garbled description of Sumba's past explained to us earlier that day, when we stopped to ask for directions using Verhoeven's map, showed that such accounts often bore little resemblance to objective historical reality. Could oral traditions and folklore ever retain any veracity about ancient historical events? And might this unconventional source of information provide novel insights about past biodiversity, too?

Indigenous communities around the globe possess a huge body of knowledge about local biodiversity, natural resources and environmental conditions. Intellectuals have sought out this Indigenous knowledge for as long as there has been interest in the natural world. Aristotle used the popular wisdom of local fishermen to inform the *Historia Animalium*, his hugely influential classical treatise on natural history, and Enlightenment-era naturalists such as Joseph Banks collected local accounts of animals and herbal remedies during their travels. Chinese scholars, such as the Qing Dynasty intellectuals Fang Yizhi and Nie Huang, also consulted fishermen and other locals to learn about unfamiliar wildlife.[5] Such information was often gathered informally through methods also used to collect folklore by scholars of the day, such as the Chinese writer Pu Songling, who 'is said to have gathered tales by the roadside, offering travellers cups of tea and pipes of tobacco in exchange for their stories.'[6] However, as early as 1802, French naturalist François Péron adopted a more formal systematic approach, using a 33-point questionnaire to interview a sealer named Daniel Cooper about a species of dwarf emu found only on King Island to the north of Tasmania. The reason that Péron used this novel approach was because he could find barely any dwarf emus himself (although during his time on King Island he was fed emu eggs and emu stew, which he considered 'truly exquisite'). The King Island dwarf emu is yet another sorry example of a remarkable island species that vanished rapidly following the onslaught of humans; Cooper told Péron that he had personally caught or killed more than 300 of them, and they were gone within a few years of Péron's visit.[7]

These data-gathering approaches are now used widely by researchers working with traditional communities to understand local awareness about biodiversity. Indigenous knowledge is an invaluable source of information about local medicinal and other practical uses of plants, and has generated the entire academic discipline of ethnobotany. It's also regularly used to establish new baselines about the status of cryptic species that might otherwise prove difficult for researchers to track down in the field. Indigenous knowledge can provide crucial insights into when and why now-extinct species disappeared, and has guided the rediscovery of other species that were thought to be extinct. It has even led to new discoveries of scientifically unknown animals and plants, from the famous discovery of the okapi by Sir Harry Johnston based on local accounts of a native 'donkey' called the *atti*, to more recent finds such as long-legged tree crabs in India.[8] Fringe researchers suggest that this technically counts as cryptozoology. And there are numerous examples of Indigenous knowledge fuelling important discoveries close to Sumba. Europeans were first alerted to the existence of Komodo dragons when local stories of a 'land crocodile' living on a nearby island reached Lieutenant Jacques van Steyn van Hensbroek, who was stationed on Flores with the Dutch colonial administration in 1910. More recently, Gregory Forth's ethnographic research into local zoological knowledge in rural communities on Flores has suggested that scientifically undocumented freshwater turtles and coconut crabs (the world's largest terrestrial arthropods) might occur on the island.[9]

The use of Indigenous knowledge to inform ecology and conservation is associated with the widespread assumption that traditional systems of ethnotaxonomy – the diversity and identity of species that are 'ethno-known' in folk zoologies – show close correspondence with scientific classification, as a result of local peoples' intimate knowledge and deep understanding of nature. This viewpoint was expressed as early as 1839 by the South Seas missionary linguist David Cargill, who considered that the Fijian language 'furnishes a name for every tree, shrub and species of grass which the islands produce ... each species of yams, talo and banana, is distinguished by a name peculiar to itself ... A name is found for every fish which frequents the shores of Feejee'.[10] Another nineteenth-century missionary in New Zealand, William Colenso, described the Māori language as 'having proper names for every natural thing however small'.[11] In the mid twentieth century, the influential French anthropologist Claude Lévi-Strauss challenged the widespread opinion that so-called 'primitive' peoples were only capable of 'mystical' participation in nature, and instead considered that members of such societies tried to understand the natural world through logical classification of structured symbolic relationships, due to a will or inherent instinct for disinterested knowledge. This reinterpretation of 'primitive thinking' formed the basis of structural anthropology, which aimed to identify the fundamental similarities in human thought across societies that superficially seemed very different. The natural units of biodiversity recognised by Indigenous peoples were thus considered to correspond closely with 'real' biological species. Support for this theory came from field taxonomists such as Ernst Mayr, who conducted extensive research in New Guinea and worked closely with Indigenous Papuan communities. Mayr identified 137 different bird species in his study sites, and found that his Papuan assistants had distinct local names for 136 of these species. Mayr noted that only two dull, nondescript warbler species were regarded as the same type of bird by local people. His conclusion was that 'Stone Age man recognises the same entities of nature as Western university-trained scientists.'[12]

This paradigm of a 'Western-style' comprehension of biodiversity amongst Indigenous communities prompted speculation that their knowledge of locally-occurring animals might persist even after such species had disappeared, whether in the form of their original species concepts, or transmuted into myth or folklore. In addition to Nage tales of the ebu gogo, numerous other monsters or creatures in Indigenous stories have been interpreted as possible folk memories of now-extinct animals. European chroniclers in New Zealand recorded Māori tales of a large lizard known as *kaweau* or *kawekaweau*, which had supernatural powers and was only seen by people who were ready to join their ancestors in the spirit world. In 1987, herpetologists Aaron Bauer and Anthony Russell described a new species of seemingly extinct huge gecko from an old specimen found in a museum in Marseille, which was possibly collected in New Zealand sometime between 1833 and 1869. They suggested that this beast, which they named *Hoplodactylus delcourti*, was the origin of the Māori tales of the *kawekaweau*.[13,]* The *pouakai* or 'old glutton', a giant predatory bird from Māori folklore that would swoop down and kill humans, might be a folk memory of the now-extinct Haast's eagle, one of the largest ever birds of prey. The chickcharney, a metre-tall feathered bipedal goblin from Andros Island in the Bahamas that could allegedly turn its head round through 360 degrees, might be based on one of the extinct dustbin-sized owls of the Caribbean. And an ogre from folklore in Madagascar with the body of an animal and face of a human, which was unable to move across flat surfaces and could be rendered helpless on smooth rock outcrops, is suggested to represent one of the island's remarkable extinct sloth lemurs.[14]

These candidate species probably all died out only a few centuries ago, but claims are also made for the persistence of folklore about other animals that disappeared much further back in time. Native American legends include huge beavers, great stiff-legged beasts that cannot lie down, a beast with a fifth arm coming out of its head, and a terrifying animal called 'bladder-head boy; or, the monster that ate people'.[15] These creatures have all been suggested to represent 'the vestigial remnant of a memory from a prehistoric age' of the real giant beavers, mammoths and mastodons of North America's megafauna, which became extinct over 10,000 years ago.[16] The bipedal sloth-like mapinguari of South America is suggested to be a description of an extinct giant ground sloth, which vanished around the same time. The Gigantomachy – the catastrophic battle between giants and the gods of Olympus in Greek mythology – was interpreted by Sir Edward Burnett Tylor, the Victorian anthropologist who founded cultural anthropology, as an ancient memory of the human-caused extinction of Europe's Ice Age megafauna.[17] Stretching the temporal limits of Indigenous knowledge even further, Aboriginal stories about bunyips, a huge kangaroo called the *yamuti*, and the

---

* However, recent research suggests that the Marseille specimen was instead collected on New Caledonia, and should be placed in its own genus *Gigarcanum* rather than with other *Hoplodactylus* geckos from New Zealand – thus raising further questions around its possible connection with stories of the *kawekaweau*.

mihirung paringmal or 'thunder bird' – a giant flightless bird said to have existed when there were still active volcanoes in western Victoria,[18] and which could kill men with its vicious kicks or 'by hugging them with his large flappers'[19] – have been proposed to represent accounts of Australia's extinct megafauna, which disappeared a staggering 45,000 years ago.

There are many such stories. But the problem they share is that these proposed identifications of long-extinct species are highly speculative. As well as being shaped by their local cultural contexts, descriptions of such creatures in folklore typically provide few characteristics that can link them definitively to their supposed extinct role-models, making any association with a particular species an educated guess. And if they were inspired by real species, it's hard to rule out the more parsimonious possibility that they're just fantastical versions of animals that are still around – the living beavers, kangaroos, lizards, owls and emus that still occur in the landscapes around these Indigenous communities. Even some of the mythological Māori terror birds might just be based on the unearthly sound of the aerial display of New Zealand snipe. These tiny wading birds are sadly also now extinct across the New Zealand mainland and only survive on tiny offshore islands, but their disappearance mostly occurred within living memory rather than many centuries ago.

Maybe the most convincing supposed Indigenous recollections of now-extinct species are Māori oral traditions of large birds generally referred to as *moa* (also the Polynesian word for 'chicken'). These accounts have been widely regarded as folk memories of New Zealand's giant flightless birds, which disappeared following the arrival of the Māori around 700 years ago, and were named 'moa' by nineteenth-century zoologists on the basis of these traditions. Nineteenth-century commentators such as William Colenso and George Grey suggested that all Māori knew of moa as 'a bird well known to their ancestors',[20] or that 'there were well known (although vague) notions current among the older Maoris concerning the former existence of a gigantic bird'.[21] Māori *whakataukī* or ancestral sayings include expressions such as 'lost as the moa is lost', which has been interpreted as both a memory of moa and an Indigenous comprehension of extinction. However, although some *whakataukī* are suggested to pre-date European arrival on the basis of linguistic analysis,[22] moa became famous across New Zealand following the discovery of huge subfossil bones in the early nineteenth century, and it's possible that concepts of moa and their disappearance might have become retrospectively incorporated into Māori lore at this point. Indigenous knowledge of moa was also variable. A 1904 account describes how an old Māori man saw a moa skeleton in a museum, and 'complained that "the arm-bones were missing." He was corrected—"but the *moa* was a bird!" The old man replied, "O son, I thought the *moa* was a man."'[23] Indeed, many historically recorded Māori accounts of moa 'were mere marvels, some stating that the creature had the face of a man, that it "lived on air", that the last living specimen dwelt on a mountain peak guarded by two great lizards', and other nineteenth-century scholars considered 'if the Moa existed up to comparative recent times, the remembrance of it might nevertheless have died out amongst the Maoris, and there might be no allusions to it on their traditions'.[24]

These various problems of identification are compounded by a further issue. Contrary to what is sometimes claimed, it turns out that ethnotaxonomy is not always straightforward to interpret, and sometimes shows little correspondence with science-based taxonomy. Indeed, rural and Indigenous peoples do sometimes misinterpret things about the natural world. For example, amphibians are widely killed in rural India because they're thought to eat cardamom on plantations, and are thus regarded as the cause of substantial economic losses to local livelihoods – but research by Indian herpetologists has demonstrated that this is a complete myth.[25] As a result of this mistaken assumption, there is sadly minimal support for amphibian conservation amongst many rural Indian communities. Similarly, I would say it's highly unlikely that mongooses in Madagascar *really* kill their prey by farting down their burrows, despite what David Burney and Ramilisonina were told in Belo-sur-Mer.[26]

More fundamentally, there can be huge confusion over what 'ethnospecies' actually are and what they correspond to, in relation both to biological species and to each other. Higher-order classification of natural units by Indigenous cultures is particularly variable, and often associated with symbolic rather than biological frameworks such as social classes or parts of the body. Indeed, when the writer Jorge Luis Borges invented a fictional tenth-century Chinese classification of the natural world, the *Heavenly Emporium of Benevolent Knowledge*, it was thought to be a genuine historical document by some modern scholars,[27] suggesting that its ridiculous categories of animals aren't entirely beyond the pale. This is despite Borges's fantastical emporium including such faunal categories as 'those that at a distance resemble flies', 'those that tremble as if they were mad', and 'those that have just broken the flower vase'.[28]

Local names for animals and plants often vary from place to place, a problem already noted by Chinese scholars centuries ago.[29] Even in the UK, humble woodlice have a bewildering variety of local names across different counties, including crunchy bats, monkeypeas, cheesybobs, chisel pigs, bellybuttons and flumps; over 150 monikers are recorded for these unassuming invertebrates.[30] The number of species recognised in folk taxonomies also vary hugely, with potential for both overdifferentiation and underdifferentiation. Sometimes single biological species represent multiple ethnospecies. This was ably demonstrated by David Fleck and colleagues in their study of saki monkeys in Amazonian Peru, where the local Matses Indians unanimously recognised two distinct, non-interbreeding types of saki called *mamu* and *bëshudu*, which reportedly differed in appearance and ecology. However, the researchers could find no differences between specimens of these two ethnospecies – they clearly represented the same biological species, although the different names might possibly refer to younger and older individuals.[31] Other Indigenous communities have been shown to overdifferentiate 'folk species' of distinctive animals such as carnivores, tapirs and anteaters; indeed, multiple synonyms for large animals appear to be the rule rather than the exception in some cultures.[32] Conversely, species with less cultural, economic or practical importance, which are harder to tell apart, or are generally less obvious (such as lizards or insects), are often lumped together or omitted from local lore. This lack of discrimination was first recorded by Aristotle, who noted that local fishermen didn't have special names for smaller types of *karkinoi* (crabs); he explained this by saying that local people didn't observe nature carefully since they did not seek knowledge for its own sake[33] (in contrast to what Lévi-Strauss later thought).

Other cultural factors also influence whether species are locally recognised and defined. Cross-cultural analyses have revealed that, counter-intuitively, the ethnotaxonomies of hunter-gatherers often contain fewer biological taxa than those of agricultural societies. This pattern might reflect the need for subsistence farmers to understand wild varieties of domesticable species, for more settled societies to require more material artefacts derived from the natural world, or that periodic crop failures might make farmers more reliant on understanding wild resources in order to survive times of famine.[34] And more subtle differences in cultural viewpoints might also govern how people classify biodiversity around the world. Lévi-Strauss famously noted that 'animals are good to think with', to be interpreted metaphorically and symbolically within cultural as well as biological frameworks.[35] For instance, sinologist Roel Sterckx highlights a passage from the early twentieth-century novel *Le Tentation de l'Occident* as providing a telling insight into different ways in which westerners and Chinese people might view the natural world. Species concepts are suggested to constitute 'a means toward knowledge' for Europeans, whereas in Chinese culture they are instead 'closely connected with our sensibility'. When thinking about a cat, the imagined Chinese narrator explains that instead of seeing a cat in his mind, he visualises 'an impression of certain supple, silent movements peculiar to cats.'[36]

In extreme cases, rural communities can be almost completely unaware of the biodiversity around them. An informative example is given by Charles Woods, who conducted fieldwork during the 1970s and 1980s in Haiti, the economically deprived western part of Hispaniola. Villagers in south-western Haiti typically referred to the region's three surviving medium-sized native land animals – the hutia, the solenodon (a large venomous insectivore), and the rhinoceros iguana – using the same name 'zagouti' (a local variant of 'hutia'). This had obvious potential to create confusion, as Woods experienced:

One day a group of people arrived at his house across from the market place in the center of Jacmel with a live "zagouti". He was very excited, but when the animal was brought to him it turned out to be a large and very abused iguana. They had found it many kilometers to the east near the town of Belle Anse, and had transported it tied upside down on a long stick born [sic] on the shoulders of two men. Woods released the exhausted, but otherwise healthy creature in his courtyard, and sat down to discuss the "zagouti" with the group (which had grown to be quite a crowd by that time). "This was not a zagouti," he said, "because all zagouties have hair." "Oh yes," the spokesman for the group said, "I know that, but this journey had been so long that all of the hair has fallen off of the animal." Then Woods said, "Zagouties have ears, and you can clearly see that this creature does not." "Yes," the spokesman for the group replied, "but the journey was through thick woods, and the ears broke off." He, and the rest of the members of the group had answers for every objection Charles Woods raised. Most of their efforts and replies were designed to make him approve of their gift, and also to convince him that he should pay them for their find. So, the problem was not that they really believed their stories. The problem was that they really did not know the zagouti well enough to recognize one when they saw it, and that they jumped on any large four-legged creature that generally matched the description.[37]

There is a depressing postscript to this story. Woods ended up keeping the iguana, but one day it escaped into the capital Port-au-Prince. The newspapers reported that it was a demonic sorcerer's familiar, and was wearing dark sunglasses and looked like the Haitian dictator François Duvalier, the infamous 'Papa Doc'. A national alert was broadcast and the iguana was shot.

None of this is to say that Indigenous cultures do not recall long-extinct species. However, the differing ways that such cultures classify and interpret current biodiversity means that we cannot assume they automatically will, or even can, retain folk memory of past biodiversity. Unfortunately what little empirical evidence we have on this issue tends to support the unlikelihood of long-term recall. This is shown by my own research. The Yangtze River dolphin or baiji was declared probably extinct in 2006, and a team of Chinese colleagues and I conducted several hundred interviews in riverside fishing communities along the Yangtze to see if any local fishers (who spent much of their lives on the water) had seen baiji recently. We collected no convincing new sightings to provide any hope that the species was still around. However, I was also shocked to find that younger fishers didn't even know what type of animal we were talking about – even when we showed them photos of baiji, gave verbal descriptions, and used local names reported by older fishers from the same villages who were sitting only a few metres away. An even stronger pattern was shown for local knowledge of the Yangtze paddlefish, a spectacular freshwater beast that could reach 7m in length – it was probably the world's largest freshwater fish – but experienced a population crash in the 1980s caused by overfishing and dam construction. The paddlefish had only been documented a few times since the mid 1990s and was also probably extinct. Of the people who had started fishing in the mid 1990s or later, over twenty per cent had never heard of baiji and over seventy per cent had never heard of paddlefish; and most young people who knew about baiji had learned about it from local news or visits by government officials trying to enforce conservation measures, rather than from older family members or other fishers.[38]

Both of these distinctive, charismatic species had been culturally or economically important well within older people's lifetimes in the riverside communities we visited – the baiji or 'goddess of the Yangtze' had been the subject of local stories and legends, and paddlefish had been the focus of economically important local fisheries, with 25 tonnes harvested annually until the 1970s. But legends hadn't been enough to immortalise the baiji; as soon as these species stopped being encountered regularly, they immediately started to become forgotten. This is a striking example of a wider cultural phenomenon called 'shifting baseline syndrome', which occurs when older people don't pass on their knowledge and experience of the past. Under these conditions, what's considered 'normal' will shift with each generation, because the younger generation has no frame of reference or wider perspective to realise that change has taken place. In environmental terms, shifting baseline syndrome can lead to potentially dangerous acceptance and complacency about increasingly degraded environments. If people have never been told their village was once surrounded by forest rather than farmland, or have never heard about the bigger fish or larger catches their grandparents used to bring home, they might have less concern about sustainability or protecting natural resources – and no realisation that the erosion of biodiversity that occurred in the past might continue into the future.

Such shifting baselines of environmental awareness are widespread. As noted during nineteenth-century discussions over the veracity of Māori lore about moa, 'the disappearance of a conspicuous animal, once existing, from popular memory may be elsewhere paralleled. Thus the common brown bear was certainly not unfrequent in Britain during the Roman occupation, and probably for some time after', but nineteenth-century scholars 'did not know that there is any allusion to it in the early English writers, and so large an animal might have been expected to survive in popular talk and story, as it does in Germany to-day.'[39] Edward Lhuyd, the early keeper of Oxford's Ashmolean Museum, noted in 1695 that the Welsh name for beaver was *avangk* (possibly derived from the Celtic word for river, *avon*), but now that the species was disappearing, 'the vulgar people of our age, scarce know what creature that *Avangk* was; and therefore some have been perswaded, that 'twas a *Phantom* or *Apparition* which heretofore haunted Lakes and Rivers.'[40] It's likely that there were still beavers in a few rivers across Wales when Lhuyd wrote this, but the locals had already begun to forget what beavers even were. Rare and declining species also become interpreted as supernatural entities in other cultural contexts, as Gregory Forth notes for dugongs on Flores – such creatures cease to be readily accessible to the perception of ordinary people, and thus no longer represent ordinary animals.[41] And sometimes, creating further confusion, the names of recently-extinct species become transferred to other animals that still exist in the local landscape, making it unclear what's even being referred to in folk memory. A striking example of this phenomenon occurred following the disappearance of the dodo on Mauritius. After this iconic bird became extinct in the 1600s, the name 'dodaarsen' became used for the island's other large flightless bird, the chicken-sized red hen (a type of rail), confusing later attempts to work out exactly when dodos actually disappeared.[42] Sadly the red hen didn't survive for long either – it was equally vulnerable to human colonists and their introduced mammals, and was driven to extinction around 1700.

I've encountered 'extinction amnesia' time and again when talking with traditional communities, in situations where the phenomenon can be recognised thanks to the existence of independent records about past biodiversity. 'Out of sight, out of mind' is seemingly the rule rather than the exception. For instance, historical archives on the Chinese island of Hainan record that two large carnivores – wolf and dhole, or Asiatic wild dog – survived locally until at least the 1930s. However, although rural communities are likely to have been well aware of these animals when they were still around, when I interviewed over 700 people living near the island's last patches of native forest, nobody at all had any idea that either species had been present in the landscape almost within living memory.[43] And in Vietnam I've asked people about their knowledge of rhinos, which used to occur across much of Southeast Asia well into the twentieth century. Several people were aware that there'd once been animals called rhinos, but the concept of what these animals actually were had become eroded and distorted now the rhinos had gone. One man said that rhinos were large animals with horns on their noses … and were 300m long and lived underground.

And different types of Indigenous knowledge relating to lost species are themselves lost at different rates. In the Hainan interview survey, we were particularly interested in collecting local information about the Hainan gibbon, now the world's rarest primate. Once found hanging from trees across most of Hainan, this species is now barely hanging onto existence, with a total surviving population of only 36 animals. Old people living near the one patch of forest that still contains gibbons told us folktales about the species, such as the story of two children who were sent into the forest by their wicked stepmother to find food, and who grew long arms to reach fruit and fur to keep warm at night and thus slowly turned into gibbons. A few people elsewhere across Hainan also still remembered such folktales (showing that every community was probably once familiar with them), but generally people living near forests where gibbons had vanished several decades ago would instead only recount more practical information – the best ways to hunt gibbons, or how their arm-bones could be made into chopsticks to test for poison.[44] This striking variation in what people told us across different parts of

Hainan suggests that folktales about now-extinct species might be the first aspect of Indigenous knowledge to be lost following the extinction of a species. This might be because folktales are typically passed down within traditional communities by older people – storytellers, shamans, and respected elders – who will probably also be the first people to disappear themselves, from old age. Conversely, knowledge of how to catch and utilise animals is shared more widely across communities, in particular with younger hunters, and thus might be expected to persist for longer. Indeed, numerous traditions of how to hunt moa were recorded from Māori folklore by nineteenth-century ethnographers.[45]

The speed with which Indigenous knowledge about large charismatic animals has been eroded and lost across these traditional cultures, often within living memory of the time when such species were still around, offers little optimism that folk memory of extinct species might be resilient in other contexts. These findings thus undermine the hope that folktales about mili monggas and ebu gogos might represent accounts of ancient human encounters with long-gone island hominins. Conversely, they might also suggest that if local peoples possess traditional knowledge about unknown animals, such species might still exist somewhere within nearby landscapes, otherwise they'd have been forgotten about. It's an intriguing idea, and one that might conceivably help biologists to locate scientifically-unknown animals in the future, through working more closely with Indigenous communities and being guided by their knowledge. Whether this approach could ever lead to the discovery of real-life mystery man-beasts in a remote mountain range or tropical forest is another question, though.

So the available evidence suggests that folk memory isn't an effective repository of historical knowledge about locally-vanished animals. Can the oral traditions of non-literate communities provide a faithful record about anything else from the distant past, though, or give any indication that Sumbanese stories of unusual historical incidents involving hairy wildmen might conceivably have some basis in fact? Some scholars have suggested 'there is no backward limit in historical time to which it is not possible for the oral tradition to refer',[46] even if such traditions might tend to be 'metamorphosed by time into the universal currency of dwarfs, ghosts and goblins.'[47] However, others have doubted this proposal, viewing claims for the historical basis of folktales with considerable scepticism. In the words of Joseph Hall, Bishop of Norwich, writing about oral tradition in 1686, 'What foundation of truth can be laid upon the breath of man? How do we see the reports vary, of those things which our eyes have seen done?'[48]

The amateur anthropologist Fitzroy Richard Somerset, 4th Lord Raglan, writing in 1936, opened his influential book *The Hero: A Study in Tradition, Myth and Drama* with a memorably sweeping statement: 'Only the smallest fraction of the human race has ever acquired the habit of taking an objective view of the past'.[49] Dismissing the suggestion by historian Raymond Chambers that 'even in heathen times ... the succession of monarchs and the length of their reigns may have been committed to memory with considerable exactness',[50] Raglan highlighted what he perceived as widespread 'pedigree-faking' in oral traditions, citing instances from classical Greece to contemporary Albania and from medieval Welsh, Saxon and Icelandic epics. Instead, he regarded transmitted knowledge in non-literate communities as strictly utilitarian. In Raglan's opinion, when old land claims, blood-feuds or similar incidents dating from past generations were eventually settled, the cultural memory of these events would become lost as there was no practical reason to remember them anymore.[51] He considered that incidents not recorded in writing could only be remembered at most for about 150 years. Beyond this point in time, history was forgotten and became replaced by myth, and Raglan noted that many fantastic tales from folklore had allegedly taken place about this far back in the past.[52]

This proposed upper limit for when historical events in folk memory will blur into myth and fable is surprisingly consistent with the alleged dates for the mili mongga stories we'd heard in East Sumba – they supposedly happened five generations back, or a little over a century ago. Interestingly, the last occurrence of some other fantastic beings elsewhere in Indonesia, such as the 'unperfected people' of Tanimbar, is also reportedly around the start of the twentieth century.[53] And this date matches how far back I know anything about my own family history. I've heard only fragmentary narratives of implausible events or eccentric behaviours from a few generations ago, worn smooth from retelling. A great-aunt with a glass eye had ball lightning come down her chimney and told her husband to put it out with a wet hessian sack; a great-grandmother lived on a bus. The oldest ancestors I've heard about are my great-great-grandparents: one was Lloyd George's hairdresser; one was a village healer in Finland; one had the second sight and saw a vision of a drowning boy, and supposedly died of psychic exhaustion after rushing to save him; and one great-great-grandmother (from the branch of the family that allegedly came to England as part of a German oompah marching band) loved the Napoleonic Wars so much that she named her sons Arthur Wellington Gartenfeld and Leonard Napoleon Gartenfeld, the latter of whom was arrested for being drunk in charge of a mule. And one great-great-grandfather was a Victorian or Edwardian market gardener in the English Midlands, who used to go to Birmingham to pay the bill for the chimney soot he used as fertiliser (something that people apparently did in those days). So the story goes, he had a few drinks in Birmingham, fell asleep on the train back and missed his stop; when he woke up, he was so surprised by being at the wrong station that he had a heart attack and fell down dead on the platform. Apparently he was only identified thanks to the signature on the soot bill in his pocket. His children were sent to the workhouse because the family then lost all its money. I know nothing else about this man before his final fatal escapade – I can't even recall his name. And I definitely can't vouch for the veracity of any of these stories; but this is what has been passed down to me (or at least this is all I can remember being told, which amounts to the same thing). This loss of the knowledge and life experience of our forebears is continual and ever-present, and is mourned in our literature; in the words of Danish writer Tove Ditlevsen, 'Oh, Granny … what you've forgotten to tell me about your life will now never be revealed.'[54]

In some Indigenous cultures, even this relatively limited level of folk memory is absent. This situation is taken to a fascinating extreme by the Pirahã, an Amazonian tribe who appear to interact with the world only through immediate experience – the evidence of their own eyes and ears, or at most, the direct experiences of people they know personally. This remarkable cultural constraint means that the Pirahã language has no perfect tense and few abstract concepts, they rarely talk about events witnessed by someone they knew who is now dead, and they have no folklore or sense of history. Their cultural memory is thus seemingly defined by the duration of a *saeculum* – the length of time equivalent to the lifetime of an individual tribe member, or the renewal of the Pirahã population. A side-effect of this worldview is that missionaries have been unable to convert the Pirahã to Christianity, because this belief system is dependent upon accepting the veracity of stories for which there is no living eyewitness.[55]

However, the idea that oral traditions can recall much older events, and that folklore and myths contain a germ of truth seen through a glass darkly – that 'history keeps peeping through the fiction'[56] – is widespread. We see this in the suggestion that fairies are a half-forgotten folk memory of earlier prehistoric inhabitants of Britain, as promoted in the horror fiction of Arthur Machen. Indeed, one of Machen's characters states that 'I became convinced that much of the folk-lore of the world is but an exaggerated account of events that really happened'.[57] This 'historical theory of mythology' is known as euhemerism after the pre-Socratic Greek philosopher Euhemerus, who considered that mythology was 'history in disguise',[58] and rationalised that the tales of the Gods of Olympus were based upon accounts of real ancient rulers or heroes, distorted and exaggerated through many retellings (for instance, Zeus was supposedly an ancient king of Crete). Genuine historical antecedents have been proposed by folklorists and academics for many such traditional tales and customs. Was the story of Hansel and Gretel inspired by a great famine in the fourteenth century? Does the ending of *Sir Gawain and the Green Knight* – where Gawain is almost ceremonially beheaded in a marsh on New Year's Day – represent an echo of an ancient pre-Christian sacrificial rebirth rite, which many centuries earlier had been enacted on the Iron Age bodies found in peat bogs?[59] Are local legends associated with many British stone circles, about merry maidens and pipers being turned to stone for dancing and playing on the Sabbath, based upon ceremonial practices conducted at these sites millennia ago, as viewed from the disapproving perspective of medieval Christianity? These kinds of fascinating suggestions prompted Edward Burnett Tylor to develop his anthropological Doctrine of Survivals in the nineteenth century, which was intended to encourage other scholars to take folklore seriously:

> When a custom, an art or an opinion is fairly started in the world, disturbing influences may long affect it so slightly that it may keep its course from generation to generation, as a stream once settled in its bed will flow on for ages … an idea, the meaning of which has perished … may continue to exist, simply because it has existed.[60]

Most of these possible 'survivals' cannot be confirmed or disproved. Intriguingly, however, some claims for the continuity of truly ancient tales can seemingly be verified. One innovative approach has been to use statistical techniques developed in evolutionary biology to investigate the distribution of fairy tales across different Indo-European languages, compared against the known timings of when these languages diverged from one another. This suggests that several stories, including the well-known fairy tales 'Beauty and the Beast' and 'Rumpelstiltskin', have a 50 per cent or greater probability of tracing their origins back to the split between Celtic, Romance, Germanic and Slavic languages 6,800 years ago. Other fairy tales could be even older, such as 'The Smith and the Devil', which is suggested to have an 87 per cent probability of dating back to the origin of all Indo-European languages over 7,000 years ago.[61]

If stories might therefore persist for centuries or millennia, maybe they can indeed provide a chronicle of ancient happenings rather than just constituting works of the imagination. However, in the words of historian David Henige, the issue 'is not that oral data *cannot* be transmitted indefinitely without distortion – of course they *can*. But what are the odds that this will actually happen and, more to the point, can be demonstrated to have happened?'[62] Events are embedded within different ways of thinking; they're described and framed within culture-specific myths, metaphors and contexts that form part of a wider corpus of tradition, with such stories having specific social functions and contributing to a wider process

of defining cultural identity.[63] As such, Indigenous stories present a
challenge to outside interpretation, and their underlying themes
and meanings can be opaque even amongst the communities in
which they are spoken[64] – making it difficult to recognise whether
they might really contain a historical component. Even if stories
can endure for surprisingly long periods, they undergo constant
narrative tweaking and embellishment – both accidental and
deliberate – to bring them up to date and maintain their relevance
for each new generation of listeners (for instance, in terms of
changing moral obligations and local politics within specific
communities). As noted by Christina Thompson, oral traditions
'are actually quite present-centered ... only what matters to the
living is retained.'[65] Folklore has function: the 'past' is narrativised
and undergoes a continual process of retrospective reconstruction,
and history becomes its near-namesake, a story. As Thompson also
points out, oral histories are good on the who and why, but not the
where and when. Tribal genealogies such as the one recounted to
us by the rato in Migurumba are thus woven full of famous heroes,
supernatural and magical events, moral lessons and guidance, and
reminders of social obligations and prohibitions, community values
and group identities; but they are notoriously hard to align with
objective chronological timelines, and typically blur from the
historical into the legendary and mythological. Other local
genealogies collected by anthropologists on Sumba describe lineages
that are descended from women who copulated with goats or eloped
with swordfish. One is even called the 'urine descent line' after a
baby who peed on her mother's lap – apparently such a monumental
event that it passed into local legend.[66]

   This ongoing process of oral evolution will eventually obscure
the details of specific historical events that were once embedded
within local narratives, although how long it takes for these details
to become lost is unclear. And this is a contentious topic. Attempts
to interpret the supposed historical veracity of traditional Indigenous
stories have been clouded by misinterpretation, wishful and uncritical
thinking, and secondary cultural contamination (as seen in the
possible Māori recollections of extinct moa). The subject has even
become embroiled in politics, with debate over whether oral histories

can constitute admissible legal evidence about the deep history of Indigenous peoples.

In some cases, though, independent corroboration of the events recounted in traditional stories is available from other historical archives – revealing that sometimes these tales can seemingly preserve records of local history for a surprisingly long time. Multiple examples now exist of Indigenous stories that appear to contain accounts of known historical events from centuries earlier. In 1928, an Aboriginal Wirangu woman called Susie from Spencer Gulf sang a story of a beautiful white bird that 'came flying in from over the ocean, then slowly stopped and, having folded its wings, was tied up so that it could not get away.'[67] This is thought to describe the arrival of the HMS *Investigator*, the sailing ship commanded by Matthew Flinders on the first circumnavigation of Australia in 1802. The timings of some older events recounted in Indigenous genealogies also seem to align with independent dates available from historical or archaeological research, such as the prehistoric Polynesian migrations across the Pacific recounted in the traditional Māori account of the 'Great Fleet' (although this particular example has been the subject of extensive debate). The incredible archaeological site of Nan Madol on the Micronesian island of Pohnpei, a ruined stone city built upon artificial islets connected by a network of canals, has also been described as 'remarkable for its close correspondence between oral traditions and archaeology',[68] with the timing and pattern of the site's construction as recounted in local folklore matching findings from excavations and carbon dating.

New discoveries about the famous stone circle of Stonehenge also provide unexpected credibility for an ancient story. Geoffrey of Monmouth's *History of the Kings of Britain* – an embellished synthesis of disparate older pseudohistorical traditions, written in the twelfth century and one of the most popular books of the Middle Ages – claims the circle once stood in Ireland. It was supposedly dismantled and shipped to Salisbury Plain by Merlin and an army who defeated the Irish and wanted the stones for their magical healing properties, and as a memorial for the war between Britons and Saxons. Remarkably, in 2021 the remains of an ancient dismantled stone circle were reported at Waun Mawn in the Preseli Hills of south-west Wales, with an identical diameter to that of the circular ditch around Stonehenge, with shapes and imprints matching those on the bluestones in Salisbury Plain, and also oriented towards the midsummer solstice. It's thought that parts of Stonehenge originally stood at Waun Mawn and were physically moved to southern England.[69] Although Geoffrey of Monmouth got some details wrong – the stones came from Wales instead of Ireland, and were moved around 3,000 BC, long before arrival of the Saxons – it seems that folk memory of Stonehenge being transported from the west was preserved in local tales for thousands of years.

Most famously, the Trojan War that forms the backdrop to the events of the *Iliad* is apparently based upon a real Bronze Age conflict involving the Mycenaean Greeks, as evidenced by Schliemann's archaeological rediscovery of Troy in the nineteenth century, although scholars continue to debate the relationship between fact and fiction within Homer's epic poem. Another famous European oral epic, the *Kalevala* of the Finns (compiled from old oral traditions in the nineteenth century), is suggested to record a striking prehistoric event of a different kind. When the gap-toothed hag of the north steals the sun and moon, the sky-god Ukko gives a spark to one of the maids of heaven to craft new heavenly bodies; but she carelessly drops the flame, which tumbles from the sky somewhere beyond the River Neva and burns many lands. A group of craters in western Estonia – situated beyond the Neva from the geographic perspective of the Finnish *Kalevala* poets – have been found to represent the remains of a meteor impact that occurred a few thousand years ago during the Nordic Bronze Age. The impact is estimated to have caused a catastrophic explosion comparable in magnitude to the bombing of Hiroshima; excavations have revealed charcoal from forest wildfires extending at least 6km from the epicentre of the blast, and the disappearance of local farming or human habitation for about a century.[70] This shocking event appears to have left such an impression that its memory was preserved in local folklore for millennia.

The *Kalevala* example is far from unique as a possible folk record of an ancient environmental catastrophe. For instance, several Indigenous stories have been interpreted as plausible accounts of ancient volcanic eruptions known to have occurred thousands of years ago, such as the

Native American Klamath account of the origin of Crater Lake in Oregon. This story, describing a conflict between the Above-World and Below-World chiefs that caused a mountain to collapse, leaving a hollow that filled with water, is consistent with geological evidence for the implosion of Mount Mazama over 7,000 years ago.[71] The oldest seemingly verifiable oral traditions are those of Aboriginal Australians. Indigenous stories from communities living close to the sea recall a time when the coastline used to be much further out, and describe the land becoming inundated by rising seas, often catastrophically. Some accounts are straightforward descriptions of the former existence of dry land in places now under the sea, whereas others are embedded within myth; for instance, the Narrangga tribe tell of when Spencer Gulf in South Australia was once a wide valley filled with marshes and lagoons, but a giant kangaroo used a magic bone to cut a trench, allowing the sea to break through and flood the valley. Today these landscapes are all covered by shallow seas only a few tens of metres deep, and were all exposed above water when sea levels were lowered during Ice Age cycles. Geographer Patrick Nunn and linguist Nicholas Reid have identified 21 groups of Aboriginal stories about coastal drowning from all around Australia, and consider them compelling because 'they tell essentially the same story, yet one that is specific to a particular coastal geography.'[72] Based on known timings of past sea-level change, Nunn and Reid suggest the stories plausibly represent folk memories of real flooding events that occurred over 7,000 years ago, and possibly as long ago as the Late Pleistocene. This apparently widespread survival of ancient accounts of dramatic environmental change, across numerous Aboriginal communities, led Nunn and Reid to propose that deep history was more likely to survive in oral traditions of non-literate communities that were culturally isolated, had a strong attachment to their landscape, and placed particular importance on teaching traditional knowledge – for example, to provide accurate recall of how their ancestors managed to survive during hard times.

In light of such studies, there has been increasing interest amongst many researchers to identify environmental or celestial events in traditional Indigenous cosmologies and interpret them as faithfully-recorded factual accounts of ancient oral history. This 'ethnogeological' or 'geomythological' approach is taken to its extreme by the suggestion that Aboriginal stories of eruptions at Budj Bim in South Australia might represent eyewitness testimony passed down for over 1,220 generations, since the last eruption of this volcano almost 37,000 years ago.[73] However, other commentators are more critical. David Henige suggests that the Klamath story 'reeks of Biblical feedback',[74] and notes that it isn't recorded in any early ethnographic accounts of folklore from the Crater Lake region. Many local tales of fiery objects falling from the sky, which describe the origin of meteor craters, could also have been influenced by recent scientific interest in these sites.[75] Such stories, as well as Indigenous accounts of prehistoric volcanic activity, may instead be better viewed as more recent attempts to understand the formation of unusual landscapes – much like the many Indigenous interpretations of unusual fossils – rather than preserved 'static narratives' representing ancient eyewitness accounts.[76] Other stories, such as Native American oral traditions interpreted as evidence for an exploding comet that led to the decline of the Hopewell culture of the Ohio River Valley, have instead been challenged as simply too ambiguous to regard as definite historical events.[77]

And independent evidence suggests that catastrophic environmental events are sometimes forgotten incredibly rapidly by local people, much like the former occurrence of charismatic large animals. On the morning of 30 June 1908, a massive explosion took place near the Stony Tunguska River north of Lake Baikal in Siberia. Thought to have been caused by a meteoroid entering the atmosphere and exploding a few miles above the Earth's surface, the explosion flattened trees across an area of almost 1,000 square miles, and caused bright atmospheric disturbances that allowed people as far away as Paris to read newspapers unaided at night. It was 'perhaps the most memorable event in recorded history'. And yet, when Soviet researchers visited the region in the 1920s to investigate what had happened, they found that local testimonies were often vague and lacking in detail – 'in a comparatively short time the fall of the meteorite was almost forgotten and memories had to be revived fifteen years later.'[78]

So these conflicting opinions make it hard to tell whether stories about past encounters with mili monggas – or ebu gogos, or Celtic fairies, or other magical or supernatural beings – might represent ancient cultural memories of long-lost hominins, or historical ethnic groups, or whatever else we might propose as speculative explanations. Although folk memories of locally extinct species don't seem to last for long, maybe oral traditions of other environmental events might persist for centuries or even millennia under some circumstances – but this is not guaranteed.

The Dutch colonial administrators considered, somewhat dismissively, that 'the Sumbanese do not know much about their own history.'[79] However, Sumba is a remote and culturally conservative place, the only remaining island in the region where the majority of rural people, living in traditional farming communities and recognising ancestral tribal affiliations, still follow their pre-colonial religious practices and beliefs. These characteristics might promote long-term survival of oral history from deep time, and there's evidence that Sumbanese oral traditions might indeed contain accurate reflections of prehistory. The stories recorded by Janet Hoskins about an older Indigenous people who existed on the island before the arrival of farmers – the Lombo or Karendi people – are consistent with archaeological, linguistic and genetic evidence for early colonisation by Qata hunter-gatherers and later arrival of Neolithic agriculturalists around 4,000 years ago. However, these traditions don't necessarily represent folk memories going that far back in time, since the two peoples could have coexisted on Sumba for a considerable period. But they do provide other tantalising suggestions that historical knowledge might have survived locally for millennia. The widespread story of how the ancestors of today's Sumbanese people reached the island describes how they crossed a mythical stone bridge from lands to the west. The place where the bridge reached Sumba is said to be near Wunga village, at the northernmost point of the central part of the island. Intriguingly, analysis of variation in the languages and dialects spoken across Sumba reveals that the inhabitants of Wunga and nearby villages retain within their speech the highest number of so-called 'Proto-Austronesian cognates' – words that can be traced to a common ancestor shared by multiple languages, and that appear to derive from the ancestral version of Austronesian spoken by the first Neolithic farmers to reach the island. Languages spoken elsewhere across Sumba instead contain more cognates derived from ancient Papuan languages spoken by the earlier colonists of Wallacea. This linguistic variation suggests that Wunga is indeed likely to be close to the site where Austronesians first colonised Sumba several thousand years ago. As these colonists spread across the island, they would probably have incorporated words from existing local hunter-gatherer languages into their vocabulary, especially if this demographic expansion was relatively gradual and involved extensive intermarriage between new arrivals and previous inhabitants, as the available genetic evidence suggests. Although Sumba was never connected to other islands by a real land bridge, it's possible that this story about a mythical stone bridge might represent a genuine ancient folk memory about where the first Austronesian colonists settled on the island.[80]

However, Sumba has also experienced major historical disruption to its Indigenous cultures during recent centuries, through the introduction of new weapons and diseases, and the heightened demand for slaves from outside traders and colonial powers. Any ancient oral traditions would have faced serious challenges to survival throughout this upheaval. And maybe folk memory can't ever stretch as far back as 45,000 years, if this is our guesstimate for when Wallacea's endemic hominins might have died out. If Sumba's hypothetical endemic hominin died out then and didn't survive for any longer. *If* it had existed in the first place. The idea that the mili mongga might be an ancient folk memory of an undiscovered extinct wildman was thus all rather complicated and confusing... and impossible to prove.

Whether or not local traditions could reveal much about Sumba's ancient biodiversity, these rich sources of information were certainly worth exploring to try to learn more about the island's other enigmatic wildlife. So we started asking people what they knew about giant rats. Such beasts certainly seemed to feature in the island's folklore. Near Kodi, Janet Hoskins had recorded the folktale of a girl who marries and is made pregnant by a handsome man who turns out to be a rat (a literal rat, not just a love-rat), and tries to take her back to his village beneath the ground; but when she can't fit through the hole, he kills her and cuts her up, dragging her underground piece by piece. After four days she is reincarnated as a new rice crop. Another version of this story sees the girl sacrificed by her brothers, with her grieving suitor turning into a rat so he can accompany her underground.[81] 'Grandfather Rat' is even one of the suite of part human, part animal Marapu deities whose offspring supposedly founded Sumba's different clans,[82] and rats represent clan totems on the island.[83] Rats and mice might also actually be witches wandering abroad at night.[84]

We were told similar stories by people we spoke to, such as the initially exciting but then somewhat rambling tale of a girl who was kidnapped by a giant rat – which turned into a long debate over the wedding dowry between the rat and the girl's relatives, and fizzled out when the respective parties couldn't come to a financial agreement, so the rat let the girl go. Might these stories constitute folk memories of extinct giant rats – or, given the rapid loss of local knowledge about extinct species seen elsewhere in the world, might they even suggest that giant rats were still alive somewhere on Sumba? But nobody thought there were any particularly large or unusual rats on the island now; forest rats were called *kalau uomang,* and a few people in the villages we visited had vague ideas that some were black and others might be a bit paler brown, maybe, but they were all about the same size – just what you'd expect for an island now overrun by various lineages of invasive rodents. The only time we'd heard about allegedly giant rats anywhere on Sumba had been the animals from the bat cave back in Mahaniwa, but the decomposing rat soup that Tim retrieved from the cave had just been the remnants of a normal black rat. I resigned myself to the most likely explanation – the protagonists in these stories were just exaggerated versions of the invasive rats found everywhere across Sumba, rather than anything to do with the extinct giant rats we'd found in Liang Lawuala. Ah well.

But if Indigenous knowledge wasn't getting us very far in understanding the fate of Sumba's giant rats, at least we could keep looking for fossils. Another obscure paper I'd found reported the existence of fossil wood on 'Soemba', collected in 1939 as surface finds from three localities near Waikabubak.[85] The wood was thought to have been preserved by burial under volcanic ash, as this part of Sumba also contained the geological remains of ancient volcanoes; it was presumably several million years old. If we could find these fossil wood beds, might there also be a chance of finding ancient animal remains that pre-dated the Quaternary bones we'd discovered so far? Was this a chance to look further back into Sumba's evolutionary past?

Umbu had heard about fossil wood in a river south of Waikabubak, and we organised a local guide to take us there. The river was 'very close', he said. We were briefly interrupted by a man from a nearby hut who told us he'd unearthed a fossil monkey skull, but it turned out to be an old statue. Two dogs, little more than skin and bone, tried to muster up the energy to either sniff us or bark, and gave up on doing either.

We set off along a muddy path, enveloped by clouds of incredibly yellow butterflies and cascades of birdsong. We walked. And walked. Every now and again we passed tiny groups of huts; people would rush out, grab our arms and eagerly thrust chunks of fossil wood caked in betel spit at us. An old man with a red betel-stained mouth and a ragged green football top talked excitedly with Umbu. 'Pak Sam! He says there is a giant fossil coconut nearby! Do you want to see it?'

How far away was this river? We'd been walking for hours… Umbu talked more with the old man. 'Oh, the river's a few miles away. And it's up a mountain. It's very hard to get to.' I decided to re-evaluate this little excursion. 'But what about the giant coconut Pak Sam?' Umbu turned to me. 'You need to have a banana, so you don't have a heart attack!' His face broke into a huge grin. 'We need more bananas!'

We were exhausted and starving when we eventually got back to the car – no river, no fossils, no bananas, and not even a giant coconut. Our guide chatted with Umbu. 'Pak Sam! He says that fifteen years ago, people from the village found some huge bones in the hills, where they go to dig gravel. There was a head and some ribs, and the ribs were huge – they were four fingers in width.' The guide gestured to us with his fingers. 'They buried them again somewhere near the village.' The guide tried to phone someone for more details. 'Ah, there is no reception!' He continued chatting while we drove back to the village. Some people had found scrap metal that came from Noah's Ark, but they weren't sure which part of the ship it was from … a monkey called Bongu that was as big as a man had visited his grandmother once for dinner … and then there had been that giant whale … his voice droned on. No one in the village knew where the giant bones were buried, and we were all tired. We headed back to

Waikabubak. Christian rock was blaring through the dusk when we arrived at the hotel; and as we walked through the dimly lit lobby, a familiar skeletal figure whispered at me from the shadows. 'Massage…?' I dodged away up the stairs.

We continued trying to find the sites marked on Verhoeven's dodgy treasure map. We spent days hiking through landscapes of jagged coral boulders, pushing past vines heavy with evil-looking brown pods that were covered with thick stinging hairs. The heat was unrelenting and exhausting. The local guides told us to place offerings of betel nut and a few rupiahs at the mouth of each cave we entered, otherwise the spirits wouldn't reveal anything sacred to us; but all we found was bat guano and modern dog and horse bones. In one cave, a huge nauseating mat of fungus stretched a web of tendrils across the floor. A child's t-shirt lay half-digested under the spreading fungal mat. Thorny vines tangled around my huge unwieldy hiking boots and tripped me as I tried to clamber over boulders. It was too hot to think; I followed blankly behind the others as though I was walking in my sleep. In the words of seventeenth-century scholar John Aubrey, 'This searching after Antiquities is a wearisome task'.[86]

One afternoon we followed a path through the trees that climbed a steep slope above a narrow valley. The village below seemed to be near one of the places named on Verhoeven's map. We were followed by most of the village children, who were laughing and fascinated by us strange foreigners. The trail ended at a low cliff rising over a small clearing, an ideal place to sit and look across the forest we'd just traversed. The top of the cliff formed an overhang that sheltered the ground beneath to create a dry, spacious rockshelter. It was almost cosy. This felt like a perfect spot for prehistoric hunter-gatherers to set up camp, and the forests below could have provided all the food they needed. I started excavating a shallow test pit in the dusty dry sediment against the wall of the rockshelter. All of a sudden, it was chaos – the village children swarmed around us, scooping out dirt with their hands and enthusiastically handing me twigs and stones and anything else they could find. Even Umbu couldn't make them stop; I tried a desperate mixture of handwaving, frowning and smiling in a vain attempt to calm the situation and prevent the uncontrolled digging.

And amongst the confusion, Paul called out to me excitedly. He was holding an ancient stone tool, teardrop-shaped with sharp cutting edges running from either side of its point. It nestled perfectly in the palm of the hand, and was made from a harder type of rock than the limestone around us – it had been painstakingly fashioned elsewhere and brought here by an unknown prehistoric hunter, who must have sat exactly where we were crouched now. Big grins spread across both of our faces. Within moments, our shallow pits yielded more tools, made of stone and of sharpened seashell transported inland from the coast – Verhoeven's *Muschelschalartefakten*! – and also buried fragments of ancient pottery. Then Umbu handed something else over to me. It was the jawbone of a *Raksasamys*, immediately recognisable by its huge bulging teeth; and next to it, a leg bone of a giant rat. And the leg bone was burnt.

It was a prehistoric treasure trove. Even without any carbon dates, these distinctive artefacts told a clear story. The shell tools and pottery were characteristic of the Neolithic Austronesian-speaking culture that reached Wallacea 4,000 years ago, close in time to the last-occurrence dates we had for the giant rats. A few thousand years in the past, a Neolithic hunter had sat right here looking out over the trees, and had cooked and eaten a giant rat caught from the forest below. Maybe the rats had been easy to catch, or maybe they'd been ferocious like the giant rats on Flores; maybe these Neolithic hunters even had special words to describe the way *Raksasamys* fought back when cornered, like the Nage have today for *Papagomys*. We'll never know. But what these ancient bones told us with certainty was that Sumba's giant rats had been hunted for food. Was this hunting sustainable, or did it lead to overexploitation of naïve, tasty island mammals? Maybe forest burning and the introduction of invasive rodents hadn't wiped out these remarkable animals – had hunting alone been enough to push them over the edge? We can probably never be sure; but hunting wouldn't have helped. Maybe the giant rats never stood a chance.

We made our way back down the hill as night was falling, stopping outside our guide's hut while the stars began to come out. The high metal roof loomed above us like a long black shadow. I sat and drank strong coffee that his wife brought out, sieving the grains between my teeth. The sound of distant gongs drummed through the darkness. Our guide talked about the mili mongga – it was a giant or *raksasa* that lived on top of a nearby volcanic hill, in a clearing surrounded by forest. We'd driven past the hill earlier today; I hadn't seen any mili monggas on it when I'd glanced idly from the car window, but I suppose I hadn't been paying that much attention.

When we reached the village at the bottom of the hill, a wedding reception was in full swing at the headman's house. Katy Perry was blaring at full volume, and someone had put a plastic chair on the big grave outside the front door. The headman, his face flushed, insisted we call him Papa. He pulled each of us towards him as we entered and rubbed his bright red nose against ours, insisting this was a 'Sumba nose kiss'. Papa held court. He talked at us merrily about magic stones and spells that could make knives stand on end, periodically breaking his rambling monologue to force us to stand up and high-five him. In between the high-fives, he kept leaning over and scratching Paul's arm sharply with his forefinger. Paul sat politely, looking puzzled and slightly pained. I forced a smile; I wanted nothing more than to get back to the hotel and wash off the cave filth I was caked in.

Tim leaned forward. 'Papa, we'd like to show you what we found today on the hill. Please can you make sure that people from the village don't go back up there and dig more holes after we've gone? There isn't any gold or anything valuable to sell, but the objects we found are really important for understanding the history of Sumba. Could you make sure it stays undisturbed?' He pulled out the two best stone tools we'd found, wrapped carefully and placed in resealable plastic bags for protection. 'Look – these are prehistoric tools from thousands of years ago, that were buried on the hill.'

Papa reached out and took the two tools from their bags. 'For fire?' he asked.

'No, probably for cutting meat,' Tim explained.

'No,' Papa insisted. 'They're for making fire. Look!' Before we could do anything, Papa held the two tools up and smashed them hard against each other. *THWACK!*

There was a moment of stunned silence. Tim hurriedly reached forward and grabbed the damaged tools, stuffing them back into his bag away from further injury. Umbu tried desperately to smooth things over and keep the conversation going. 'Papa, do you know anything about the mili mongga?' That bloody mili mongga – !!

Papa's face broke into a broad grin; he was oblivious to what he'd just done. 'Yes, I can tell you all about that,' he boomed. 'There is a grave behind one of the houses down the road, which contains the bones of a giant. The shin bone is more than a metre long.' He demonstrated the size of the giant's leg by whipping his bare foot up and slamming it down on the table in front of me, right into a bowl of rice. 'How do people know there's a giant buried inside – has anyone dug it up?' I asked. 'The story has been passed down from their grandparents,' Papa intoned.

The sound of beating drums carried in from the tropical night outside. This was something different from the wedding celebrations. 'The Marapu ceremony is starting,' Papa said. 'They are getting ready to move some bones from another grave outside the kampung – it's rude for people to leave their relatives buried outside the village. It's expensive to move the bones, so it doesn't happen often. First, you have to ask the spirits if they're happy to be moved. You scatter some ash on the ground, and the spirits will show their sign by making animal footprints in the ash. Then there will be a second sign; a fireball will come down and make another mark in the ash.'

The door opened and drunken guests from the wedding poured in, talking quickly and loudly. One of the new arrivals produced two long yellow balloons full of palm wine, as though he was about to make a sausage dog at a children's party; instead, he emptied the boozy water balloons into cups and forced us to drink. It tasted of fruity whisky. The rato in charge of the bone-moving ceremony stared at me, unsmiling, his mouth smeared red with betel nut. I felt extremely tired; nothing seemed quite real any more.

'The mili mongga bones buried here are sacred,' Papa continued. 'The grave can be opened with just two pieces of thread, but it would be taboo.' His eyes glittered. 'You have to pay me a hundred million rupiah if you want to look at the bones. Even if you want me to tell you about the giant, you will have to pay for a ceremony with a pig with big tusks.' He gestured drunkenly with his hands to show how big the tusks would have to be. I kept a slightly glazed polite expression on my face.

At last it was time to leave. We stood up for a final pained high-five and nose kiss, and one more scratch for Paul for good measure. I fought to stay awake during the long drive back through the darkness. We'd found another piece of the puzzle about the giant rats, but the solution to the mili mongga mystery remained stubbornly elusive. If only we could take a look inside one of the graves everyone kept talking about, I thought vaguely, as I was jolted awake again by the bumps in the road. But would that solve the mystery, or just create a whole new one?

CHAPTER TEN

THEY MIGHT BE GIANTS

Sad, alas, the man who dreamt of Fairies!
For a single dream spoiled his whole life.

Bai Juyi

We continued to prospect Sumba's arid landscape for fossils, tramping around under the burning sun through coral rubble and thorns. Our guide knew all the local caves because he visited them to collect the nests of swiftlets to sell. We sat on rocks under the shade of a large tree, drinking water and eating the eggs and hot chillies we'd brought for lunch, before lowering ourselves into the darkness of yet another cave. It was very wet underground, with big pools of mud in the thick black sediment. At least we were sheltered from the smothering heat outside. At the back of the chamber, one of the teenagers who'd accompanied us from the village was playing with what looked like a pair of sticks. I realised he was tapping on a human skull with a pair of shin bones. Other parts of the skeleton stuck out from the black mud all around us.

As Tim gently took the drumsticks away from the teenager, I examined the scattered bones around us. It was the old encrusted skeleton of a modern human. 'It's probably someone who used to live around here a long time ago,' I said, as we sat in the mud around the dirty bones. But the others started debating excitedly. 'No, Pak Sam,' said Umbu. 'It's not anyone from the village. These guys say he was probably Japanese. Or maybe it's one of the first Portuguese to come to Sumba hundreds of years ago.' They were adamant.

Surely the skeleton was all that was left of one of the hundreds of past generations of villagers who had lived nearby, and who had somehow ended up here under the mud. Why insist that these bones represented something else, one of only a handful of foreigners to have ever set foot in this remote corner of the island? Did the huge importance of observing the correct funeral customs on Sumba mean that a body which wasn't properly buried could not be perceived as part of the local community? Or did the response to this skeleton reveal something about how other people on Sumba might have interpreted other old bones? Was this a clue to the identity of the ultimate 'other' on Sumba, the mili mongga?

Back outside the cave, I had another look for the endemic buttonquail in the thin dry grassland as the sun started to sink into the sea. An Es Krim seller pedalled slowly across the empty landscape far below, the faint repetitive bang of his gong drifting up and becoming lost in the harsh blue sky. The buttonquail eluded me again. As I gazed out over the limestone terraces stretching down towards the sunset, the air filled with bats and glowed pink and gold in the unreal evening light, as though the dusk was made of soft fire.

We drove from village to village, asking about fossils, caves, and mili monggas. The hours rolled by as we bumped along endless dusty roads. Little stalls with triangular corrugated roofs looked like roadside shrines, but sold glass coke bottles full of yellow petrol. I stared out of the window at tiny Sumbanese horses galloping across rolling grassy hills. I still couldn't get used to the betel ceremony whenever we visited each village; it felt weird to start hoicking up huge red wads of saliva as soon as I met new people, and I couldn't aim my betel spit accurately. It didn't emerge from my mouth as a nice neat jet, no matter how I tried to purse my lips – instead it went everywhere down my chin and over my notebook. I sat trying to be polite while huge volumes of red fluid poured out of my mouth, my mind blank from the effects of the betel nut.

Many of the people we spoke to had heard of the mili mongga. Sometimes they called it a tiger because of its furry body, and naughty children were afraid of it. Maybe their grandparents' grandparents might have seen it, or knew of people that had, and maybe it was real once upon a time, but it didn't exist any more. And other people told different stories.

'The mili mongga is short, maybe a metre or less in height. It looks like an orangutan, but with long fur that's grey, like a monkey's. Even though it's small, it has big power – it can even make trees fall down. About ten years ago a woman from our village went to catch fish in a river that flows through the forest, and a mili mongga grabbed her but she managed to run away. Maybe it wanted to eat her, or communicate with her, although in the old stories mili monggas can only talk to each other but not to people. Or maybe it wanted to mate with her? She's dead now, anyway. It's the same sort of thing as an orangutan, but it can stand and walk like a human. I don't know if it's a person or an animal, but it's not a spirit. Although, maybe it is a sort of spirit, because not everyone can see it.'

'There are two kinds of these things. The meu rumba is large and has big breasts, and it eats people. The mili mongga is small and eats forest snails. The mili mongga is not the child of the meu rumba – they're different things. My family have seen both of them. A meu rumba came to my grandmother's house when she was about 15 or 20 and had just got married. It eats humans and was looking for food. It was completely covered with greyish white hair as long as my finger, and didn't wear any clothes. The meu rumba is afraid of dogs; when it went into my grandmother's house, the house platform broke and it fell down onto a dog, and the dog started barking so it ran away. It was three metres tall. The mili mongga doesn't wear any clothes either. My father met a mili mongga in the forest – he just saw it and watched it playing around. There were two of them, maybe a metre or a bit shorter, about eighty centimetres tall; they were playing in the forest, but they ran away and hid when they realised that someone had seen them. This happened when my father was about twenty. I will be ninety-five next month, so this happened maybe a hundred years ago, during the Dutch period. He told me the mili mongga has white pale fur, but not long like that of the meu rumba – more rough, like goat fur. It has feet and hands like a human. Its head is a bit different to a man's, a bit more like a dog's. It walks on two legs. You can tell males and females apart by their nipples. Female mili monggas have long nipples.'

Other people told us even more stories. In one village, our host told us about a huge footprint that was 10m long, made by a giant that walked past two or three hundred years ago. We sat on the porch beneath old rusa deer antlers mounted on the outside wall.

The sun shone down onto the breeze-block houses at the edge of the village, and white clouds floated high above the water buffalos bathing in a muddy pool beside the rice paddy. The scarecrows in the paddy were made of bottles tied to pieces of string on bamboo poles; they had no effect on the flocks of tiny munia birds flying continuously back and forth over the heads of the golden rice stalks. 'Something else happened around here when I was a child. One summer about fifty years ago, a strange man dressed entirely in white arrived in the village. He looked like one of you foreigners!' our host laughed. 'He was just blown in by the wind. As soon as he arrived, everyone heard the sound of a flute, and they all hid inside because they thought he'd come to cut people's heads off. Other villages have the same story – they call him the moron gali – and it's happened in Sumatra too. This all happened several times that same year. I didn't actually see him myself,' our host admitted. 'But I did hear the sound of the flute.' He pursed his lips and made a peeping noise. 'Two men followed the moron gali; one was from this village. Later, these men said they'd walked around, looking for people whose heads they could cut off. But everyone was afraid of them so they weren't punished. Neither of the men said they ever saw the stranger's flute either. That's another mystery.' Then the old man started recounting another story he'd heard as a child about a giant bird that helped the villagers know when it was time for the harvest.

And an ancient old man dressed in a skater hoodie told us the mili mongga was strong and tall and lives in wild places, such as forests that contain big rocks, but nobody has ever actually seen it, not even its bones or footprints. 'If you go to remote villages, people will tell you not to go into the forest because of the mili mongga; but it's just a word, not something that's actually real. It's all just stories. Since Christianity came to Sumba, people think it's the same thing as Satan. It's just another forest spirit, like the patau patuna. You've not heard about that? It's a sacred sound. If you're walking through the forest, you might be able to hear it; you have to be very quiet. The patau patuna is a person… and it's also like a wall, which you have to cross. If you encounter the patau patuna, you can hear singing and talking, but you can't understand what it's

saying. I've heard it myself in the evenings. It's a sign — you have to
return home straight away if you ever hear it. The patau patuna is a
ghost that can communicate with humans, like the ghosts that live
in the large trees and rocks in the forest, the patau tana. When it's
in the rocks and the big trees it's patau tana, and when it's in the
forest it's the mili mongga.' The conversation then drifted onto the
subject of Satan, about whom the old man had strong opinions that
he wished to share.

We had a different reason for visiting Lambanapu, a village south
of Waingapu that overlooked a broad curve of river. When I went
down to see if any fossils were eroding out of the sediments on the
foreshore, a huge monitor lizard skittered down into the water and
swam strongly across the current in sinuous curves. Umbu told me
to watch out for crocodiles. Further up the slope at the edge of the
village, near a pig sleeping happily under a megalithic tomb, was an
area of uneven soil that had been dug over. Back in Java at the start
of our trip, between the Brandon's Flaming Avocados and the
kedongdong juice, we'd visited Indonesia's National Centre for
Archaeology to talk about hobbits and stegodons and fieldwork on
Sumba. One of the archaeologists had shown us pictures of a dig
conducted a few years earlier at Lambanapu, which had uncovered
anatomically modern human skeletons from the past couple of
thousand years.

Umbu chatted with the villagers who had gathered to meet us. A
gang of children edged as close as they dared, then ran away laughing
and screaming. A tethered zebu wandered slowly amongst the gaggle
of people, forcing me to keep stepping back and forth over its rope to
avoid tripping into the piles of zebu dung all around me. 'Pak Sam!'
Umbu said. 'This old man says the bones that were dug up here were
not... *normal*.' A middle-aged woman leaning against a bamboo
pigpen agreed, grabbing a long stick to demonstrate how large the
bones had been. There was general agreement that these had been no
ordinary human bones — they had been huge. 'We all saw the bones.
They were from giants!'

First the insistence that the calcified skeleton couldn't be a local,
and now this. I *knew* that the human bones excavated here were
ordinary; we had only wanted to visit Lambanapu to get a feel for

where previous researchers had found ancient remains. Why was everyone so insistent that the bones they'd seen hadn't been human? Across Sumba we'd been told bizarre stories about buried giants' bones; but here at least I knew they were mistaken. What did this mean about all the other tales of unfeasibly large skeletons we'd heard across the island? I had a feeling that the pieces of the puzzle were finally falling into place.

It's not just in Sumba that old bones from ancient sites have been viewed as the exhumed remains of giants. As we've seen, elephant and mammoth fossils were often misinterpreted by early observers as the bones of a bygone race of huge people, and sometimes even reburied in human graves. But tales of giants and their bones are extremely widespread, suggesting there was more to this cultural phenomenon than the occasional fossil elephant. Various ancient human bones were alleged to have unnaturally large proportions during the early years of archaeology, such as the Neolithic 'Giant of Castelnau' from southern France. 'Skeletons of gigantick stature' were reported by grave-robbers and early antiquarians in ancient barrows and other prehistoric sites across England and Scotland, including 'the bones of an excessively big carcase' at Flowers Barrow promontory fort in Dorset, and a human skull unearthed by a crofter in North Uist supposedly 'so vast that when placed on his own head it covered his shoulders'.[1] Tradition also tells that St Michael's Mount in Cornwall was constructed by two giants, Cormoran and Cormelian, and when the medieval church on the Mount was excavated in 1720, a skeleton nearly eight feet tall was supposedly 'found walled up in a cellar, together with a leather water jug, still as good as new'.[2] The belief that ancient barrows contained the remains of giants was a 'fairly common story' across Britain, with the association between barrows and giants dating back at least to Saxon times.[3]

These reports form part of a wider body of magical lore about such ancient sites, which are also commonly associated with ghosts, fairies, buried treasure guarded by dragons or curses, Arthurian traditions, and many other supernatural and mystical properties. Indeed, bones allegedly belonging to King Arthur, discovered at Glastonbury Abbey in 1191,[*] were described by Gerald of Wales as having gigantic proportions:

> ... the bones of Arthur ... were so huge that his shank-bone, when placed against the tallest man in the place, reached a good three inches above his knee ... the eye-socket was a good palm in width.[4]

But such 'giant skeletons', if they remained available for scientific study, were always shown to be human after all upon later examination (those that weren't misidentified mammoths, anyway). A few individuals had genuinely suffered from gigantism, such as a large skeleton from the third century AD found near Rome in 1991, which showed bone damage at the base of the skull consistent with a pituitary tumour that would have caused acromegaly.[5] Most just turned out to be normal human skeletons though, their size misinterpreted by the common human tendency for exaggeration, and by basic anatomical errors such as measuring the entire length of the femur when estimating height, rather than understanding how the articulating femoral head slots into the pelvis below the top of the bone.

Ancient human remains, especially those found in unusual situations, have probably inspired tales of giant beings throughout history. For instance, it's suggested that bog bodies – remarkably well-preserved (but normally-proportioned) ancient human bodies often representing Iron Age sacrifices, found in peat bogs across northern Europe – might have inspired local legends of boggarts and bogeymen, and even the terrible monster Grendel and his mother, who live beneath the waters of a boggy lake in *Beowulf*.[6] The common belief in ancient giants also reflects the consensus view of human history in the West, as maintained by Old Testament scripture and many scholars. There were giants in the earth in those days, we are told in Genesis – a statement interpreted literally until only a few centuries ago (although some considered that these 'were giants only

---

[*] The 'discovery' of Arthur's grave might have been a medieval publicity stunt to raise funds to rebuild Glastonbury Abbey after a disastrous fire in 1184.

in daring wickedness'[7]). The idea of a former antediluvian world populated by giants and purged of 'fallen angels and their monstrous children' by the Deluge[8] was promoted by influential Renaissance writers such as Annius of Viterbo and François Rabelais, who wrote a pentalogy of novels in the sixteenth century about the famous giants Gargantua and Pantagruel. Rather than fossil elephant bones inspiring beliefs in giants, ancient human remains excavated from prehistoric sites might already have been expected to be elephantine in size.

These views were also championed by Italian Enlightenment philosopher Giambattista Vico, who provided an original explanation for why giants were once common:

> *Urine and feces, Vico noted, have a great fertilizing power, as anyone can see by planting a field where an army has made camp. The Jews were clean, as their divine law made them. But pagan babies, abandoned by their mothers, played with their own excretions – as Tacitus showed, in a passing remark in his account of the ancient Germans. Pagan babies, accordingly, never stopped growing.[9]*

In Vico's words:

> *Mothers, like beasts, must merely have nursed their babies, let them wallow naked in their own filth, and abandoned them for good as soon as they were weaned. And these children, who had to wallow in their own filth, whose nitrous salts richly fertilized the fields, and who had to exert themselves to penetrate the great forest, grown extremely dense from the flood, would flex and contract their muscles in these exertions, and thus absorb nitrous salts into their bodies in greater abundance. They would be quite without that fear of gods, fathers, and teachers which chills and benumbs even the most exuberant in childhood. They must therefore have grown robust, vigorous, excessively big in brawn and bone, to the point of becoming giants.[10]*

In 1718, a French scholar called Henrion even estimated the heights of people mentioned in the Bible, and proposed that the two first humans were truly gigantic – Adam was 123 feet 9 inches tall, and Eve was well over 118 feet tall. However, 'thereafter degeneration was rapid', and Noah was only 27 feet, Abraham was 20 feet and Moses was a meagre 13 feet.[11] Some early Rabbinical scholars went even further, maintaining that Adam's head 'overtopped the atmosphere, and that he touched the Arctic Pole with one hand, and the Antarctic with the other.'[12]

Local traditions enforced these ideas. Geoffrey of Monmouth's *History of the Kings of Britain* described how Britain was first settled by Brutus the Trojan and his followers, who found the country 'uninhabited except for a few giants' that they killed or drove into caves in the mountains.[13] And across the countryside, local communities commemorated their own special giants in folktale and myth. Some even had their own tombs or graves, such as Piers Shonks the dragon-slayer in Brent Pelham and Jack O'Legs in Hertfordshire.[14]

'Giants are common to all nations, both ancient and modern', and 'most nations had the belief that the men who preceded them were of immense stature', observed Edward Wood in 1868 in his detailed compendium on the subject, *Giants and Dwarfs*.[15] Early European explorers reported giants in exotic foreign lands such as Patagonia, where Antonio Pigafetta said the head of one of Magellan's crewmen would only reach the waist of the region's huge inhabitants. Tales of a giant race of Patagonians persisted in Europe until the eighteenth century and were championed by contemporary thinkers such as Lord Monboddo. Belief in ancient giants persists in pseudoscientific fringe scholarship. For instance, Russian ophthalmologist and esoteric writer Ernst Muldashev proposed that when giants' remains were placed in sarcophagi, they 'dematerialized and turned into a kind of energy blobs',[16] offering a convenient explanation for the absence of giant skeletons in archaeological excavations.

Comparable giant myths form part of folklore from Mexico and Peru to Australia and Iceland; in the words of the Icelandic seeress who recounts the Norse creation myth in the *Poetic Edda*, 'I remember giants born early in time, long ago'. Many Indigenous communities across Wallacea, including the Nage of Flores, also describe large physical size as a general feature of their distant forebears.[17] The earliest ancestors of the Nage people of Ndora 'were uniformly big and tall. There can be little doubt that the large size of early ancestors expresses their social 'stature', their importance in the Nage scheme of things.' Some Nage ancestors 'were so large that when they urinated the flow was strong enough to carry away piglets, puppy dogs and chickens'. Others were not true giants but possessed extraordinarily large body parts, such as feet, ears or armpits. 'One ancestor is credited with an exceptionally long penis, so long indeed that, while bathing in a stream, he was able to have intercourse with female bathers considerable distances away.'[18]

Giants represent powerful elemental forces that exist in the remote wilderness. They are responsible for creating the world in Indigenous cosmologies, or a genius loci evoked to explain weather, earthquakes and landslides, or the topology of the landscape. Their names are linked to mountains, crags and imposing ruins, reflecting 'man's innate reverence for the colossal'.[19] Their menacing size and raw power symbolises the savage ancestral state, the uncontrolled and uncontrollable; they have 'impulsive morality' in the words of E. M. Forster. They're often depicted as an early unsuccessful attempt to create humans. For Jung, they represent archetypal emotional energies, the overpowering surge of feelings that 'magnifies everything in our surroundings'[20] ... 'In the psychical world, however, giants do still exist. It is completely beside the point whether they really exist'.[21] In mythology they are fought by heroes and must be eradicated before civilised society can begin. They are often also portrayed as foolish figures of fun, and are typically defeated when their low intelligence is trumped by human ingenuity and wiles, such as when Odysseus deceived the Cyclops Polyphemus – or the villagers of Miurumba tricked the mili monggas into eating corn mixed with burning pebbles.

Hairy humanoids, too, are widespread as fabulous liminal figures in story and legend. The wildman or woodwose of medieval European folklore and Arthurian romance, a 'strange relative of Homo sapiens', is interpreted as 'a lively and sometimes pungent commentary on the bestial side of his nature', giving 'external expression and symbolically valid form to the impulses of reckless physical self-assertion which are hidden in all of us, but are normally kept under control.' This creature, the folkloric sibling to Pan and the fauns of classical mythology, is an unfettered child of nature, 'compounded of intransigence, lust, and violence … a negative ideal in all its harshness and one-sidedness.'[22] Simon Schama described it as representing the antithesis of civilised Christians during the Middle Ages,[23] and it has been referred to in more tongue-in-cheek academic terms as the 'Sasquatch pastoral'.[24] G. K. Chesterton put it best: 'he has no name, and all true tales of him are blotted out; yet he walks behind us in every forest path and wakes within us when the wind wakes at night. He is the origins—he is the man in the forest.'

Woodwoses share much in common with other ambiguous, animalistic Anglo-Saxon and medieval trickster figures such as the exiled outlaws or 'wolf's heads' of fenland and greenwood, with their superhuman strength and animal companions, who 'offered an imaginary alternative to human life' and lost their humanity by shifting from men to predators to hunted beasts to extensions of the trees themselves. Did they 'represent the fears and the desires of a people in a state of constant negotiation with the land they inhabited'?[25] These entities exist in 'inverted worlds',[26] providing distorted reflections of our civilised values; they are socially constructed beings, that help to define us by showing us what we are not. Or maybe we should just appreciate their otherness and mythic potency, rather than trying to associate them with specific psychological categories or meanings. As Jorge Luis Borges said about dragons, 'there is something in the dragon's image that fits man's imagination, and this accounts for the dragon's appearance in different places and periods.'[27]

And maybe it's misleading to think of mili monggas primarily as giants. A wide pantheon of semi-human beings exists in folklore, but folklorists have long noted that their attributes and actions often become mingled. In the words of Canon John MacCulloch in 1921:

Everyone who has studied the various sets of beings, more or less supernatural, in which humanity has believed, is aware that a large number of characteristics is common to all. They have their own personality and name, they are quite distinct from each other, yet many things attributed to one set are attributed to others. So much so that it would almost seem as if, from very far-distant times, a stock of incidents existed which could be assigned indifferently to various

*denizens of the world of fancy, just as certain stories are told, now of this, now
of the other, outstanding personality.*[28]

Some Nage stories about ebu gogos have seemingly been shaped
through cultural exchange between islands and contain folklore
elements from elsewhere in Asia, and Janet Hoskins documents wider
similarities in other folktales told across different parts of Wallacea,
including Sumba.[29] Mili mongga tales also contain themes common
to mythology from around the world, such as sexual couplings with
not-quite-human entities, the negative consequences of eating food
offered by other beings, or tricking monsters by substituting stones
for food – tropes found in European fairy and folk traditions, Greek
myths, and Indigenous stories from North America to China.
Comparable story elements are seen in other accounts of wildmen;
for example, many ancient Chinese descriptions portray clearly
imaginary creatures with fantastical abilities and supernatural
powers.[30] Although Nage descriptions of the ebu gogo have been
interpreted as potentially plausible folk memories of *Homo floresiensis*,
some aspects of their description are also clearly fantastical and unlike
any possible prehistoric hominin or living primate, instead providing
a link to wider folklore. One Nage story also tells of people killing
ebu gogos by making them eat hot stones,[31] and the implausibly long
breasts that female ebu gogos supposedly carry over their shoulders
are a common characteristic of many supernatural entities. One of
the Nage's mythical ancestors is called Kedho Long Breasts,[32] and
enormous dangling breasts are also shared by the Wewe Gombel of
Java, the mythical Lady Trieu of Vietnam, and giants from ancient
Chinese stories.[33] Joseph Campbell's famous work on comparative
mythology, *The Hero with a Thousand Faces*, discusses the wild women
of the Russian forest who 'fling their breasts over their shoulders
when they run and when they nurse their children' (and who also
tickle people to death);[34] and the Faengge or Fankke, a female ogre
from the folklore of the Tyrol and Bavarian Alps, has similar
appendages.[35] 'Giantess throws her breasts over her shoulders'
constitutes an entire recognised category within the folktale
classification system of Antti Aarne, Stith Thompson and Hans-Jörg
Uther. These creatures join the pantheon of human–like entities in

folklore with characteristics that are slightly off-kilter or askew, such as a hole in their back or feet that point backwards. Such shared attributes support Lévi-Strauss's structuralist theory that fundamental commonalities underpin world mythology.

Indeed, some stories told in far-distant places bear a remarkably close correspondence to Wallacean wildman tales, with important implications for interpreting what they might mean. In particular, the Kachari people of Nagaland in north-east India talk of an 'uncanny race' of small, dark-skinned magicians called the Siemi, who lived in the jungle. The Kachari regarded the Siemi as a nuisance and decided to wipe them out, and the last few Siemi hid inside a cave. 'At last, however, the Kacharis tracked them down. They did not attempt to attack the cave, a deep, defensible tunnel. Instead, they cut great armfuls of scrub and brushwood, filled up the cave-mouth with it, and fired the whole. So the last of the jungle-people perished, suffocated, in their last retreat.'[36] The Kachari tale of the destruction of the Siemi is seemingly identical to the Nage tale of the destruction of the ebu gogo – raising serious questions about whether the ebu gogo narrative could therefore really describe a local historical event. Is trapping tiny wildmen in caves and burning them to death actually just a common theme in folklore across tropical Asia?

Hairy wildmen in other cultural traditions, such as the woodwoses of medieval Europe, are variably depicted as either giants or dwarfs.[37] Some stories we'd heard also depicted mili monggas as tiny playful forest beings rather than giants, closer to the Nage description of the ebu gogo, and somewhat like fairies in western folk traditions. This fairy connection offers further insight on how to potentially interpret these entities. Most mili mongga stories referred to events that supposedly occurred a few generations earlier, close to Lord Raglan's speculative upper limit for when folk memory starts blurring into myth. Many people specifically said mili monggas no longer existed, or had retreated into the most remote forests where they couldn't be found. Although others maintained that they still inhabited the landscape, their continued presence was only revealed by piles of green snail shells and other subtle signs. Even the old lady at Maoramba who initially told us she'd seen a mili mongga, at the start of our first trip to Sumba, had actually been talking about someone from her village who sounded suspiciously human. Mili monggas visited people long ago, with their memory kept alive by folktales and ancient tombs. They symbolised something from the past; back into storyland giants have fled.

This is also the fate of fairies and little folk, which have been consistently relegated to the past by contemporary observers through the ages. In Kipling's Edwardian tales, Puck of Pook's Hill was the last of the Old Things and the People of the Hills that formerly inhabited the English countryside. In the previous century, Jane Eyre mused that 'for as to the elves, having sought them in vain among foxglove leaves and bells, under mushrooms and beneath the ground-ivy mantling old wall-nooks, I had at length made up my mind to the sad truth, that they were all gone out of England to some savage country where the woods were wilder and thicker, and the population more scant'[38] ... "The men in green all forsook England a hundred years ago,' said I ... 'And not even in Hay Lane, or the fields about it, could you find a trace of them. I don't think either summer or harvest, or winter moon, will ever shine on their revels more".[39] Charlotte Brontë's nanny Tabby, who grew up in the 1700s, thought factories had driven the fairies away.[40] Others said they had been frightened off by electricity or cars, or by priests and prayers. Thomas Keightley wrote in 1828 that 'the sounds of the cotton-mill, the steam-engine, and, more than all, the whistle of the railway train, more powerful than any exorcists, have banished, or soon will banish, the fairy tribes from all their accustomed haunts'. Earlier still, in the seventeenth century, John Aubrey considered that 'the divine art of Printing and Gunpowder have frightened away Robin-Goodfellow and the Fayries',[41] and as long ago as Chaucer, the Wife of Bath talked of fairies in similar terms: 'This was the olde opinion, as I rede; I speke of manye hundred yeres ago. But now kan no man se none elves mo'. And such traditions persisted well into the twentieth century in places such as the Isle of Man, where locals talked of the 'li'l people', and recounted that they 'don't like the noise of church bells and they don't like the noise of trains' – they were said to have fled behind the mountains 'to get away from the noise.'[42]

Folklorist Katharine Briggs noted that fairies have always been supposed to belong to the last generation and are lost to the present one, and titled her book *The Vanishing People* to emphasise this integral feature of fairy tradition. In the words of another folklorist, fairy stories are characterised by 'keeping ... always the freshness of being associated with the near past, the day before yesterday.'[43] This represents part of a wider nostalgia for the past that has existed throughout history.[44] But, as Yeats said, 'the faery and ghost kingdom is more stubborn than men dream of. It will perhaps be always going and never gone.'[45] Rumours of the fairies' demise have been greatly exaggerated through the centuries, with sightings continuing to be reported with surprising regularity. My favourite report is given by Arthur Conan Doyle in his sadly credulous account of the famous Cottingley Fairy photos. Conan Doyle mentions a sighting of 'a dozen or more small people, about two feet in height, in bright clothes and with radiant faces', who danced 'hither and thither ... in sheer joy' on the lawn, and even used a croquet hoop as a horizontal bar. This all took place in the garden of 'the late Mr. Turvey, of Bournemouth', described by Conan Doyle as 'one of the most gifted clairvoyants in England'.[46] I am hopeful that this man might have been a distant relative of mine. However, sadly my family's oral traditions are strangely silent on the matter of fairy stories.

Whether we give any credence to such reports, we see that giants, fairies, mili monggas and other supernatural beings share similar characteristics – they are all distanced from 'real' people in folktale and myth, both by their unusual appearance and by existing only within history. They constitute mythological creatures in mythological time, figurative rather than literal entities from an imagined past, which 'has a transcendent and unquestioned character that keeps it walled off from all subsequent times'.[47]

Or maybe priests and prayers did more than just drive off the fairies; they might have helped create them. Many Victorian scholars thought fairy folklore was the last vestige of an ancient religion, either 'decayed pagan gods or local English deities.'[48] Indeed, 'it is generally accepted that the god of one religion becomes the Devil of that which replaces it.'[49] The pagan concept of wildness 'implied everything that eluded Christian norms and the established framework of Christian society',[50] and fairyland became 'a contested site in the struggle between the official and unofficial cultures of the Middle Ages'[51] – so the medieval church infernalised fairyland and suppressed wildman cults, including pagan rituals such as death–rebirth fertility ceremonies or re-enactments of the Wild Hunt. Fairies and wildmen became transformed into demons that attended the witches' sabbath. This shift in attitude is shown clearly by a seventeenth-century annotation in an early edition of Spenser's *The Faerie Queene*: 'fairies are devils, and therefore fairyland must be the devils' land'.[52] One of the last vestiges of these former rites is apparently preserved today in the chequered costume of the harlequin, originally derived from furry wildman suits worn across Europe centuries ago.[53]

Did Christianity have a similar effect when it spread across Indonesia during the colonial period? Was the mili mongga a figure from Indigenous Sumbanese animism that became bastardised through outside misinterpretation? Or is this missing the point? Maybe the most famous wildman of all, the yeti, can provide further insights. Instead of a surviving *Gigantopithecus*, many people in the Himalayas believe that yetis are actually spiritual entities; they are widely thought to be the bodyguards of Chenrezig, the deity that embodies the compassion of all Buddhas. Yeti relics kept by monasteries, such as the mummified finger smuggled to England by James Stewart, are thus considered to provide a crucial link to the world of gods and demons within this spiritual belief system, rather than mere evidence of supposed flesh-and-blood cryptids championed by western explorers.[54] Within the Sumbanese worldview, might mili monggas therefore also represent a 'hierophany' – in the words of historian and philosopher Mircea Eliade, a manifestation of the sacred, or a breakthrough of the supernatural into the world? We may see the universe as fundamentally rational and following immutable natural laws, but to others it remains an enchanted place. As Christopher Hadley wrote in his fascinating investigation of the mythical English dragon-slayer Piers Shonks, 'Searching for a kernel of truth by trying to remove the legendary elements misses something, it gets rid of the best bits.'[55] Even amongst researchers, there is

increasing recognition that 'anthropology should always be open to
the possibility of wonder.'[56] It is imperative to consider the mystery
of the mili mongga not just from our perspective as outside observers
(the so-called *etic* perspective in anthropology), but also from the
perspective of the culture that holds this differing worldview (the
emic perspective). This can be extremely difficult – we are all brought
up within our own specific cultures, with their own explicit and
implicit conventions, assumptions and prejudices about structuring
experience and making sense of the world. But if we can gain a
different perspective, we might receive some truly surprising insights
into how other cultures think about reality.

Our time on Sumba was running out once more. As we made the
long drive to another village along another dirt road, Umbu passed
the time by telling us about a masseur in Waingapu who could restore
broken bones by massaging them with a hammer. We stopped at a
crossroads to buy food and *sirih pinang*; a crowd of women crushed
frantically against the car, thrusting bowls of eggs through the open
windows and peering in at us. I cracked a boiled egg against the side
of my kneecap as we started off again, and laboriously tried to pick off
the little bits of shell into my hand while the car jolted back and forth
across the road.

We drank coffee on the porch with Umbu's friend when we
eventually reached the next village, and chatted about bones, caves
and mili monggas with the small group of people who drifted over to
meet us. What appeared to be the largest wasps in the world flew back
and forth under the corrugated roof. Our conversation was punctuated
by the rhythmic, hypnotic double *thud thud* of a runner banging on a
loom, as the woman sitting near us on the concrete floor slowly made
ikat. She unstrapped herself from the loom and went in to make more
coffee. As we dodged the huge shiny black wasps that buzzed
worryingly close to our faces, Umbu joked about different ways to be
nakal. 'And if a gecko calls after a person says something, that means
they're talking shit!'

A chicken climbed onto the ikat loom as an ancient man with faint
blue tattoos on his arms shuffled slowly over to the porch and accepted
a cigarette and a coffee. He clutched a vial of lime and his mouth was

stuffed with betel nut; he reached into a faded floral cloth bag for some chewing tobacco, which he stuffed into his mouth too. He lit his cigarette. 'The meu rumba is a type of human – you can touch it – like a huge cannibal. But the mili mongga is magical. A long time ago, people tried to learn magic from the mili mongga. A lot of people could use magic in the past; they had special power, so you couldn't cut them with a knife. If you know the secret polite word for iron and know how to talk to it in the proper way, then you can break it easily and it can't cut you.'

We gave the old man a lift to the next village, where we sat around on green plastic garden chairs outside his friend's hut. Our new host, wearing a floaty shapeless batik shirt, leaned against the breeze-block wall of the hut; one of his wives inspected us from the doorway with disinterest and wandered back indoors. The concrete yard was strewn with wet playing cards, and a weary old mange-ridden dog was sitting in the middle of the circle of plastic chairs, slowly sliding over while laboriously trying to scratch itself. A terrible squealing suddenly started inside the hut and an enormous black pig burst out of the door. I tried to smile pleasantly as the pig ran across the wet playing cards straight towards me. The conversation continued when the laughter died down. 'If somebody meets a mili mongga, they have to go back home and taste some salt,' our new host explained.[*] 'If the salt doesn't taste salty, they will get sick and die. If it's still salty they will live to an old age. This is because the mili mongga has magical powers.'

'Does anything else have similar magical powers?' I asked.

Our host leaned forward conspiratorially. 'Yes – GOATS. There is a special type of goat called ngau; it's alive, but it's also a spirit, because not everyone can find it. You can request something from it, but it's not easy – you have to fight it. And you will get sick if you see a ngau. When I was a child, I saw a ngau! I hurried home and tried some salt,

---

[*] Gregory Forth also reports this local tradition on Sumba: 'According to another common belief, not everyone is spiritually suited to observe mili mongga. Thus, if someone should see one, he or she should place salt on the tongue; if the salt can be tasted, the person will survive, but if it is tasteless, death will surely follow'.[57]

but it was still salty, so I was OK.' Another pig started screaming inside the hut.

'Has anybody ever found the bones of a mili mongga?' I asked. A small man in a purple t-shirt, one of the crowd who had gathered around us, introduced himself as Matius and spoke up. 'About a year ago I found an open tomb in the forest that someone had already dug up to look for gold. There was a huge skeleton! I tried to put the skull on my head as a helmet, but it was too big! And the shin bone was as big as an entire leg. It was like a human skeleton, but huge. I left it all in the forest. I'll take you there if you like and show you the giant bones.' Umbu and I looked at each other, grins forming on our faces. 'It's in the old abandoned village in the forest.'

Finally! A grave with giant bones that had already been dug up, and that someone could actually show us! All we had to do was go and take a look. 'Let's go tomorrow! After church!' We thanked everyone profusely, offered round more cigarettes and *sirih pinang*, and made plans for the following day. All of a sudden, was the answer to the mili mongga mystery within our grasp?

We stood beside the open tomb to which Matius and Yakobus had brought us and looked down at the scattered bones, left lying here exposed on the stony coral dirt and dead leaves by whoever had originally tried to dig up the tomb. A femur. A tibia, its distal end broken off. Other leg and arm bones, all pitted and weathered and stained, their surfaces flaking and darkened. What was left of a jaw, still with a few teeth in its sockets and its right ramus intact; and something that first looked like a dirty broken plate or a bleached shard of coconut – the top of a skull. A thick black millipede crawled over the grey, weathered bones; brown earth still stuck to them in patches.

Taking another swig of warm, tasteless water from my bottle, I bent down. I guess part of me was pleasantly surprised there was anything here at all; at least this hadn't been a completely futile trek. But I wasn't really surprised by what we'd found. A symbolic shaft of sunlight broke through the canopy overhead, lighting up the gloomy forest floor and the bones lying all over it. They were completely normal human bones.

Umbu frowned with disappointment. 'I'm sorry, Pak Sam. We have still not found mili mongga bones!' Matius had a subdued expression. He avoided Umbu's gaze. 'Last year there were giant bones here, as well as this skeleton,' he insisted quietly. 'But I can't find them now. They have gone.'

Somewhere nearby in the forest, a gecko called.

CHAPTER ELEVEN

THE PERFECT ISLAND – A FAIRY TALE FOR BIOLOGISTS

Beautiful is what we see.
More beautiful is what we understand.
Most beautiful is what we do not comprehend.

Nicolas Steno

So that's what I learned about the mili mongga. But I only spent a few months on Sumba; of course I haven't solved the mystery. The people I spoke to interpreted the world around them, their history and their past, in a very different way to me. The challenge is to step out of my anthropologically etic perspective and try to understand *why* they view the world in such a different way.

Maybe, though, I've been looking at the mystery of the mili mongga the wrong way round. I tried to identify the origin of the weird stories we kept hearing – what was behind them? Were they inspired by late-surviving island hominins, by memories of former island peoples or local events, by odd fossil bones or ancient animal burials? Given this bewildering diversity of different possible starting points, how can we ever know for sure? As Errol Morris noted in a recent critique of the history and philosophy of science, 'One of the oddities of history is that legends often supersede facts.'[1] The way in which these stories first started is almost certainly unknowable – although it's probably linked to our species' propensity to conjure up imaginary giants and other monsters in cultures around the world. So is it possible that mili monggas weren't based on anything at all ... were they just stories all along? Elizabeth Pisani described Sumba as a place where 'It's not uncommon to see, silhouetted against the evening sky, a man on a stocky pony holding a lance, for all the world like Don Quixote.'[2] Following this analogy, were Sumba's giants just windmills after all?

But. If the mili mongga existed in local folklore for whatever reason, then this crazy variety of possible explanations – escaped criminals, sperm whale teeth, stegodon fossils, even normal human bones found in unusual contexts – could all help to perpetuate stories about hairy giants that live in the woods. So the mili mongga might be a kind of 'catch-all' that's used to explain all sorts of unrelated things. Weird bones in a cave? It's a mili mongga. Don't remember who built this big old wall? Must have been a mili mongga. It's hard to disprove such claims in a rural society with no written records. So, irrespective of how the stories first started, it's easy to see how they could have been reinforced and adapted to take on a life of their own. And the stories we were told about mili monggas are just part of a much broader catalogue of wild tales from Sumba's rich body of folklore and myth. If you want to crack the mystery of the local wildman, it's imperative not to think about wildman stories in isolation, without trying to understand their cultural context alongside other local narratives and traditions, fantastical or otherwise.

And of course, this raises further questions. Do these conclusions apply only to the mili mongga, or are they also a useful way to think about accounts of other mysterious critters told in different places and cultures? Other explorers seeking enigmatic cryptids have sometimes reached similar views. Speculating on the true nature of the yeti, the famous mountaineer Reinhold Messner suggested that explorers and locals 'have talked past each other for over a century, their divergent views … drifting ever further apart.'[3] Messner's conclusions provide further food for thought. He considered that the yeti legend was inspired by sightings of bears, but thought that the mystery could only be solved through attempts 'to connect two completely different modes of perception.'[4] This is an important reminder that other cultures view the world around them in fundamentally different ways.

The stories people told me in Sumba revealed how nature was seen as a network of spirits, of which mili monggas are just one among many perplexingly unfamiliar supernatural beings. William Seabrook, writing about the beliefs of the people of a different island, considered that such worldviews seem to 'lie in a baffling category on the ragged edge of things which are beyond either superstition or reason.'[5] We might think that the West has moved beyond such a viewpoint. The 'primitive' mythical imagination of Europeans in centuries past – who lived in a world ruled by magic and prayer, inhabited by spirits and by a bestiary of woodwoses, Marco Polo's dog-headed men and other fantastic creatures – was supposedly swept away by rationalism. From the seventeenth century onward and into the Age of Enlightenment, the intellectual changes associated with Europe's scientific, philosophical and technological revolutions led to a new worldview underpinned by immutable natural laws and the need for evidence, enforced by the better living standards that developed alongside and in response to this new way of thinking.[6] The sociologist Max Weber famously defined this process as 'disenchantment', highlighting it as a key step in the development of modern western society.

But it's not as straightforward as that. Western culture might appear rational in outlook, but a surprisingly large number of us also believe in the existence of another reality: a liminal, magical world of ghostly hauntings, UFOs, fairies and other paranormal events – and even some truly bizarre phenomena, such as a supernatural talking mongoose called Gef who allegedly lived in a farmhouse on the Isle of Man in the 1930s.[7] The modern superstition is that we're free of superstition[8] – or in the words of sociologist Bruno Latour, 'we have never been modern'.[9] Why is this? As T. S. Eliot said, is it because 'humankind cannot bear very much reality'? Whatever the reason, are these widespread western beliefs really that different from folktales about a hairy bogeyman that used to live in the woods?

Logic and evidence can be slippery fish, too. Science often comes down to a delicate trade-off between probabilities and possibilities rather than anything nicely clear-cut – which can provide intellectual fascination, but also frustration. Carbon dates represent probability densities rather than exact dates; the debate over exactly when to define a distinct island population as an endemic species will probably continue forever; and a bewildering range of statistical analyses is now available to try to tease patterns out from noisy ecological or social science data. And even direct evidence can be interpreted in a surprising variety of ways. To most of us, it seems obvious that fossils represent vestiges of ancient life and past worlds; or that there is an obvious distinction between humans and other primates, but their morphological similarities are indicative of shared evolutionary ancestry. But as we've seen, even though people have found fossils and interacted with biodiversity for millennia, such interpretations are surprisingly recent. 'Damned data' such as seashells on mountaintops didn't automatically prompt the acceptance of new intellectual paradigms; instead, they were usually slotted quite comfortably into pre-existing non-scientific frameworks of how to interpret the world.[10] Just like the challenge of adopting an emic outlook and appreciating local viewpoints in anthropological research today, people throughout history usually found it difficult to step outside their prevailing cultural worldviews too. In a sense, this is the true history of science, and of the big ideas about who we are and where we came from. Given this history, who can predict how much our ideas about the world around us will change further in the future? In the words of one influential anthropologist, 'He is a rash man who is dogmatic about an ancient bone.'[11]

Our research on Sumba was much more modest. But this is how science typically progresses – by chipping away slowly at the daunting edifice of the unknown, often having to make use of unexpected sources of data; and advancing our knowledge by another stepping stone or two before passing the baton onto future investigators. And the past is a foreign country. Given the vagaries of the fossil record and oral traditions, we will almost certainly never gain a full understanding of the history of biodiversity on Sumba or anywhere – gaps will always remain. When did the very last *tikus besar* finally die, bringing about the extinction of the genus *Raksasamys*? Did it meet its end at the hands of hungry prehistoric hunters; or through new diseases brought by invasive rats; or did it just perish alone in the remnants of a forest being cleared by human settlers? We'll never know this level of precision about Sumba's past. This philosophical standpoint is termed epistemological pessimism, which posits that it may not be possible to obtain complete knowledge about the world (the opposite standpoint, epistemological optimism, maintains that we can potentially understand everything, at least in theory if not currently in practice).

But if this is the case, we shouldn't despair. Compared to our baseline of what was previously known, we now have a much better understanding about Sumba's biodiversity across recent millennia. As anthropologist Janet Hoskins said about Sumba in a different context, 'This past itself was more alive and influential than I had expected'.[12] We now know that Sumba once had a similar fauna to nearby Flores – it was a bizarre wonderland of tiny elephants, Komodo dragons and other strange giant lizards, and several species of giant rats, which have all now vanished. And some of these animals disappeared only a couple of thousand years ago, pointing the finger of suspicion without doubt towards human responsibility for their extinction.

Although we found the fellowship of the hobbit – the same weird island fauna that coexisted with *Homo floresiensis* and was preserved alongside it in Liang Bua cave – we didn't find any ancient hominin remains on Sumba. Does this mean Sumba didn't have its own endemic human species? No. Flores has received considerably more palaeontological attention than Sumba, and researchers had excavated Liang Bua on and off for half a century before they found the bones of *Homo floresiensis* nearly 6m below the surface. In contrast, our work on Sumba is just the first step towards understanding its enigmatic prehistory. There is much more work to be done on this fascinating island.

Why should we care about a few old bones on a remote island on the other side of the world? In the words of Frederic Wood Jones, writing about the Cocos (Keeling) Islands – another isolated island landmass 2,500km west of Sumba – 'The drama of history loses nothing by reason of the smallness of its stage'.[13] Sumba's own unique hobbit is probably out there somewhere, waiting to be found by a future palaeontologist. It will be a monumentally exciting discovery. There might have been other endemic hobbits on Sulawesi and Timor, and maybe Sumbawa and Lombok as well. Any of the larger Wallacean islands that had unique endemic stegodons and giant rats are likely to have once been home to tiny endemic hominins too; if ancient humans made it to Flores, there's no obvious biogeographic reason why they couldn't have colonised other islands as well through chance overwater dispersal events.

The discovery of *Homo floresiensis* in 2003 forced us to think about human evolution in a radically different way, and future discoveries of other extinct dwarfed island hominins will prompt further revision of how we perceive our relationship with nature and our understanding of what we actually are. The existence of endemic human species on islands brings home the fact that we are just another animal that adapts to local environmental conditions, with an evolutionary tree that radiates and branches in response to ecological and geographic barriers just like we see in other non-human species. The hobbit reminds us forcibly that we are not divorced from the natural world; instead, we are dependent upon nature and fundamentally shaped by it. And the hobbits of Flores and elsewhere, who persisted on their isolated island homes for hundreds of millennia within long-gone tropical ecosystems, must have perceived and interpreted their world and themselves in ways that cannot be imagined. I'm just one of the eight billion living members of the sole surviving hominin species, *Homo sapiens*; but the fragmentary hobbit remains discovered so far from Wallacea are part of the much wider diversity of what it once meant – biologically, evolutionarily, and individually – to be human.

Modern humans possess diversity both within and between populations, in our genetics, physical appearance and cultures. Such diversity is the result of historical isolation of separate populations by landscape barriers such as seaways, mountains and deserts, which until recently reduced the opportunities for genetic exchange across the wide distribution of our species. Biological differences between human populations – both random genetic changes and small-scale local adaptations – have often been fixated upon for social or political reasons, but these are relatively minor. In contrast, it is the spectacular variation in our uniquely complex cultures that is a defining characteristic of *Homo sapiens* – the legacy of the 'creative explosion' that first propelled our species out of Africa and around the world.

With eight billion people alive today, there should be a huge amount of cultural diversity. Ironically though, as our population continues to grow, the world's financial, trade and media networks are becoming increasingly globalised – making us less and less culturally rich as a result. Rural and Indigenous communities around the globe were once remote from the influence of major power centres, but are now connected by transport links built to facilitate the extraction of natural resources required by ever-growing urban populations. Indigenous communities are now exposed to a homogeneous westernised 'Coca-Cola culture' that brings the allure of greater economic opportunities and material possessions compared to age-old traditional lifestyles. This cultural globalisation was brought home to me sharply a few years ago when I was doing fieldwork in the tropical rainforests of Papua New Guinea. We'd flown by propeller plane to a tiny village in the jungle with a grass landing strip, then hiked for 12 hours up a rocky river bed, spotting birds of paradise and harpy eagles in the dense forest around us. At the end of the day, exhausted from the hike, we finally reached the tiny field station that would be our base for the next few weeks – I hadn't been anywhere this remote from civilisation before. In front of the hut was a small group of smiling local men … and as we trudged up to the hut, they turned the radio on and started singing along to Lady Gaga.

It's hard to develop a measure that effectively captures the multidimensional and complex concept of cultural diversity. Maybe the best proxy is language diversity, since different cultures are typically associated with their own unique way of communicating. Each language provides a new framework for interpreting and defining the world. About 7,000 languages are spoken today, suggesting a seemingly healthy level of cultural diversity. However, a closer look at the numbers of people who speak different languages instead reveals a disconcerting story of cultural erosion. Half the world's population now speak one of only 24 'super-languages', including English, Mandarin Chinese, Spanish, Hindi and Bengali (and including Javanese as the Indonesian representative of this group). Each of these languages is spoken by tens or hundreds of millions of people, with well over a billion speakers of both English and Mandarin. At the other end of the scale, more than 5,000 of the world's 7,000 languages each have fewer than 1,000 speakers, and 3,500 languages are together spoken by a total of only about eight million people. To put it another way, half of all the languages that exist today are being kept alive by a combined population smaller than the number of people in London.[14]

And this uneven pattern isn't stable. Most of these languages – and the traditional cultures that support them – are dwindling away, their remaining speakers ageing and dying as their children and grandchildren switch to speaking one of the 'super-languages' or move away from impoverished rural communities to towns and cities in search of new opportunities. Many of the languages and their associated cultures that I've used as case studies in this book have now faded or vanished completely since their traditional knowledge was documented by past anthropologists and ethnographers. The Yukaghir of the Russian Far East, who thought that frozen permafrost mammoths were ancient animals called *xolhut*, speak a unique language family comprising two sister languages that together had only about 120 remaining speakers by the start of the twenty-first century. And the Andaman Islanders speak languages belonging to two entirely different language families, but most of these languages have disappeared completely following the arrival of European colonial powers on this island archipelago. The languages of the Great Andamanese language family are now spoken by only a handful of elderly islanders, and the last person in the world who could speak the Andamanese Sare language died in April 2020.[15]

And that's not all. The same processes of globalisation that are driving the loss of global cultural diversity are also responsible for much of the world's ongoing biodiversity loss, thanks to worldwide demand for tropical timber; the clearance of rainforests to make room for rubber and oil palm plantations and mineral extraction; the unsustainable demand for threatened wildlife species from traditional Chinese medicine, the luxury food market and other commercial drivers; and increased opportunities for accidental transport of invasive species and pathogens. So, as we move ever closer to becoming a unified 'global village', the path we're travelling is bringing about the destruction of global biodiversity, the traditional cultures that live alongside nature, and their unique ways of thinking and perceiving reality. We are in a biocultural diversity crisis.

The ongoing erosion of traditional cultures has catastrophic implications for human rights, equality and identity in marginalised and disenfranchised communities around the world. It also has catastrophic implications for biodiversity conservation. Human activities have been implicated in driving biodiversity loss for tens of thousands of years, and local interactions with nature can still be damaging and exploitative today. However, many traditional cultures have reached a sustainable equilibrium with the biodiversity they live alongside, with deep connectivity and value for the natural world as well as reliance upon natural resources. Traditional methods of managing natural resources, such as hunting restrictions or taboos based on local belief systems, can often help to maintain wildlife populations and ecosystem function. And although ethnotaxonomy and Indigenous knowledge vary hugely in both resolution and quality, this body of knowledge represents a vast repository of information about locally occurring species, their ecology and distribution, and their potential medicinal or food value – of irreplaceable value for conservation, and also potentially for wider human health and well-being. But globalisation and cultural encroachment threaten to upset the fragile balance of these social-ecological systems; traditional management is weakened by changing social pressures, the effectiveness of hunting taboos is undermined by external economic demand for valuable species, and Indigenous knowledge about biodiversity is lost as unique cultures and languages also vanish. Who knows what the Yukaghir could have told us about sustainable management of Siberian ecosystems, or what unique insights about Andamanese biodiversity and Indigenous values of nature might have disappeared forever with the words of the last Sare speaker?

Let's return one last time to Sumba. Even this remote island has been affected by globalisation in surprising ways; its forests have been under pressure from unsustainable outside demand for sandalwood for centuries, and its endemic rodents might have been wiped out by invasive rats that arrived with the first sandalwood traders. But although much of its biodiversity is gone, there is still hope.

As my travels on the trail of an imaginary wildman revealed, Sumba's rural communities still retain a rich body of traditional knowledge and lore. It would be naïve to think that all these

Indigenous cultural interactions with nature have been sustainable or provided mutual benefits for local biodiversity. The long history of landscape burning has converted nearly all of Sumba's native forests to grassland; and collecting dugongs' tears for good luck didn't do wonders for the local dugong population either. However, some of this traditional knowledge might be able to provide a community-led framework for maintaining the island's endemic species, natural resources and ecosystem services. For instance, the Marapu traditions that prohibit excessive clearance of native forests might well have saved much of Sumba's remaining tropical tree cover. These and other local values and attitudes are rich with the potential to aid conservation, if they can be better understood and integrated into regional management policies. Ultimately, the hope is that in our brave new world of mass extinction, the deep Indigenous knowledge of the people of Sumba can allow them to act as stewards for their island's vulnerable surviving biodiversity – a vision shared for Indigenous communities around the world.

I'll admit that I don't really believe in the mili mongga. So in the scheme of things, do a few local stories about a skulking wildman that eats green snails really matter? Maybe the best way to look at it is this. Whether or not stories about mili monggas are 'real', or beneficial to biodiversity in any way themselves, they constitute an integral part of the wider Sumbanese tradition of Indigenous knowledge about the natural world: how to classify it, how to interact with it and how it interacts with us, and the values and fundamental importance it holds. Maybe, therefore, we should just regard the stories of mili monggas as an indicator of the continued health of this unique body of cultural traditions, and its potential to guide the sustainable management of Sumba's remarkable social-ecological system through local community leadership.

Wallacea is the land where the world's oldest known stories are recorded, painted tens of thousands of years ago in red pigment by unknown hands. Weird humanoid figures – tiny people with snouts, beaks and tails – move among life-like depictions of pigs and buffalo along a dark cave wall in Sulawesi. Was this ancient image intended to depict a real event, or is it an entirely imagined scene? Are these curious, haunting therianthrope figures meant to be hunters, animal spirit helpers … or something even more strange? We will never know for sure. But we must hope that enigmatic, compelling tales like this will still be told long into the future across Wallacea, an archipelago so remarkable that it was once home to its own unique prehistoric human species. And to support that hope, we must work to ensure that biodiversity and human well-being can both continue to exist here on Sumba, the most fascinating island in the world.

Or maybe there really is a giant hairy wildman out there in the woods, after all.

Acknowledgements

First and foremost, I thank Umbu Palanggarimu: without him, our time on Sumba would have been completely unsuccessful. He graciously showed us around the island and explained his culture and its traditions, tirelessly contacted people on our behalf with queries about fossils, caves and mili monggas, and invited us into his home. I'm lucky to have a friend on Sumba: Umbu, thank you again. I also thank Erwin Pah, the welcoming staff of the Tanto Hotel, the people of Mahaniwa village, and everyone on Sumba who kindly shared their stories, histories and experiences with us.

To safeguard their privacy, the names of all local people mentioned in this book have been omitted or replaced by pseudonyms, and village names are only mentioned where necessary. All conversations that we had with local respondents were voluntary, and were only conducted with people aged 18 or above. We were given permission to record all of the stories that I have recounted here, based upon informed consent from all respondents following an explanation of the purpose of our visit to Sumba. All research was conducted in accordance with Indonesian national laws, under social/cultural visas sponsored by the Zoological Society of London's Indonesia Office in Bogor, and further abided by the local regulations and customs of each community we visited. All fossil material collected on Sumba during our fieldwork is now accessioned in Indonesia, in the Vertebrate Palaeontology Collection of the Bandung Geological Museum.

The adventures and discoveries recounted in this book would not have been possible without the support of many people. Special thanks go to Tim Jeffree, my constant companion in Indonesia – you helped more than you realise. I also give huge thanks to Jennifer Crees, James Hansford and Paul Barnes, who accompanied me to Sumba on different visits – although you all started off as my students,

I think of you all as close friends. Fieldwork planning was made possible by advice from Gregory Forth, Retang Wohangara, Gert van den Bergh and Anne Fortuin, and I also thank Gregory for first introducing me to the mili mongga and for conducting such thorough and considered research into the hominin-rich folklore of Southeast Asia; I hope that my pontifications on the identity of Sumba's mystery wildman are of interest. In addition to the organisations already mentioned above, invaluable logistical support and permission was also provided by Iwan Kurniawan, Erick Setiyabudi, and the staff of the Indonesian State Ministry of Research and Technology (RISTEK), the Geological Survey Institute (Bandung), the National Centre for Archaeology (Jakarta), the Culture and Tourism Department of East Nusa Tenggara (Kupang), and the Interior Ministry Office (Waingapu). Simon Young provided extremely helpful discussion and literature about fairy traditions, and in no particular order (other than alphabetical), I also thank Nick Crumpton, Kirsten Foster, Ellie Landy, Adrian Lister, Kate McClune, Tom and Kane Trezona, William Turvey, Tim Waters and Nicholas Wilkinson for further thoughts, ideas, and general encouragement and enthusiasm. Fieldwork was made possible thanks to a University Research Fellowship from the Royal Society, and my work is supported by the Institute of Zoology, Zoological Society of London. I am extremely grateful to Jim Martin and Jenny Campbell at Bloomsbury Publishing for their enthusiasm for this book and their support throughout the publication process, and Elizabeth Peters for copy-editing. Finally, of course, I thank Mum and Dad – always.

Other than specific changes made to anonymise people's identities, everything that I've described about my fieldwork on Sumba is accurate, to the best of my memory and knowledge. It is very possible that some details or nuances of the stories included in this book might have accidentally become lost in translation – if so, I apologise and take responsibility for any errors. Furthermore, I cannot vouch for the veracity of any of these stories; all I can say is, this is what I was told! In the words of Jerome K. Jerome, I hope that:

The chief beauty of this book lies not so much in its literary style, or in the extent and usefulness of the information it conveys, as in its simple truthfulness. Its pages form the record of events that really happened. All that has been done is to colour them; and, for this, no extra charge has been made.

Notes

Chapter 1: Splendid Isolation

1. Smith, B. 2017. *The Island in Imagination and Experience.* Saraband, Salford, p. 2.
2. van Duzer, C. 2004. *Floating Islands: A Global Bibliography.* Cantor Press, Los Altos Hills, CA, p. 151.
3. Ibid., pp. 90–91, 298.
4. Ibid., p. 122.
5. Hope, C. W. 1902. *Annals of Botany* 16: 513.
6. Gould, J. 1872. *Proceedings of the Zoological Society of London* 1872: 493–496.
7. Censky, E. J. *et al.* 1998. *Nature* 395: 556.
8. Nunn, P. D. 2009. *Vanished Islands and Hidden Continents of the Pacific.* University of Hawai'i Press, Honolulu, HI, p. 70.
9. Jungers, W. L. *et al.* 1997. *Proceedings of the National Academy of Sciences of the USA* 94: 11998–12001.
10. Fabre, A.-C. *et al.* 2023. *Journal of Zoology* 319: 91–98.
11. Hartstone-Rose, A. *et al.* 2020. *American Journal of Physical Anthropology* 171: 8–16.
12. van der Geer, A. *et al.* 2010. *Evolution of Island Mammals: Adaptation and Extinction of Placental Mammals on Islands.* Wiley-Blackwell, Chichester, p. 145.
13. Turvey, S. T. *et al.* 2005. *Nature* 435: 940–943.
14. Lomolino, M. V. *et al.* 2013. *Journal of Biogeography* 40: 1427–1439.
15. Köhler, M. & Moyà-Solà, S. 2009. *Proceedings of the National Academy of Sciences of the USA* 106: 20354–20358.
16. Lomolino, M. V. *et al.* 2013. *Journal of Biogeography* 40: 1427–1439.
17. Burness, G. P. *et al.* 2001. *Proceedings of the National Academy of Sciences of the USA* 98: 14518–14523.
18. Pritchard, R. E. (ed.) 2011. *Peter Mundy: Merchant Adventurer.* Bodleian Library, Oxford, p. 214.
19. Ibid., pp. 214–215.
20. Sulloway, F. J. 2009. *Journal of the History of Biology* 42: 3.
21. Lowe, P. R. 1911. *A Naturalist on Desert Islands.* Witherby & Co, London, pp. 114–115.
22. Ibid., p. 113.
23. Vázquez-Domínguez, E. *et al.* 2004. *Oryx* 38: 347–350.
24. Turvey, S. T. (ed.) 2009. *Holocene Extinctions.* Oxford University Press, Oxford, pp. 204–206.
25. Hughes, A. C. 2017. *Ecosphere* 8: e01624.

Chapter 2: Sumba, East of Java

1. Quammen, D. 1996. *The Song of the Dodo: Island Biogeography in an Age of Extinctions*. Scribner, New York, NY, p. 426.

2. Okie, J. G. & Brown, J. H. 2009. *Proceedings of the National Academy of Sciences of the USA* 106(S2): 19679–19684.

3. Wilson, E. O. 2002. *The Future of Life*. Little, Brown, London, p. 89.

4. Flannery, T. F. 1994. *The Future Eaters: An Ecological History of the Australasian Lands and People*. Reed New Holland, Sydney.

5. Earl, G. W. 1853. *Contributions to the Physical Geography of South-Eastern Asia and Australia*. Hippolyte Bailliere, London, p. 4.

6. Osipov, S. *et al.* 2021. *Communications Earth & Environment* 2: 71.

7. Wallace, A. R. 1908. In *The Darwin-Wallace Celebration held on Thursday, 1st July, 1908, by the Linnean Society of London*. Linnean Society, London, p. 6.

8. Ibid.

9. Wallace, A. R. 1869. *The Malay Archipelago*. Harper & Brothers, New York, NY, p. 160.

10. Ibid., p. 211.

11. Ibid., p. 210.

12. Ibid., pp. 25–26.

13. Stelbrink, B. *et al.* 2012. *Evolution* 66: 2252–2271.

14. Frantz, L. A. F. *et al.* 2018. *Proceedings of the Royal Society B* 285: 20172566.

15. van Welzen, P. C. *et al.* 2011. *Biological Journal of the Linnean Society* 103: 531–545.

16. Fowler, C. 2013. *Ignition Stories: Indigenous Fire Ecology in the Indo-Australian Monsoon Zone*. Carolina Academic Press, Durham, NC, p. 8.

17. Ducrocq, S. 1996. *Geological Magazine* 133: 763–766.

18. Haig, D. W. *et al.* 2017. *Palaeogeography, Palaeoclimatology, Palaeoecology* 468: 90.

19. Louys, J. *et al.* 2018. *Journal of Asia-Pacific Biodiversity* 4: 503–510.

20. van der Geer, A. A. E. *et al.* 2016. *Journal of Biogeography* 43: 1656–1666.

21. Maringer, J. & Verhoeven, T. 1979. *East and West* 29: 247.

22. van den Bergh, G. D. *et al.* 2008. *Quaternary International* 182: 16–48.

23. van den Bergh, G. D. *et al.* 2009. *Journal of Human Evolution* 57: 527–537; Hocknull, S. A. *et al.* 2009. *PLoS ONE* 4: e7241; Meijer, H. J. M. *et al.* 2010. *Journal of Biogeography* 37: 995–1006; Dennell, R. W. *et al.* 2014. *Quaternary Science Reviews* 96: 98–107.

24. Darwin, C. 1859. *On the Origin of Species by Means of Natural Selection, or the Preservation of Favoured Races in the Struggle for Life*. John Murray, London, pp. 310–311.

25. Sartono, S. 1979. *Modern Quaternary Research in Southeast Asia* 5: 57.

26. Flannery, T. 1998. *Throwim Way Leg: Adventures in the Jungles of New Guinea*. Phoenix, London, p. 158.

27. Dammerman, K. W. 1928. *Treubia* 10: 299.
28. An overview of Sumba's geological history is given in K. A. Monk *et al.* 1997. *The Ecology of Nusa Tenggara and Maluku.* Oxford University Press, Oxford, pp. 9–59.

Chapter 3: Glutton-Granny

1. Forth, G. 1981. *Rindi: An Ethnographic Study of a Traditional Domain in Eastern Sumba.* Koninklijk Instituut voor Taal-, Land- en Volkenkunde, Leiden, p. 40.
2. Shanks, G. D. *et al.* 2011. *American Journal of Epidemiology* 173: 1211–1222.
3. Maclean, C. 1977. *St Kilda: Island on the Edge of the World*, revised edition. Canongate, Edinburgh, p. 59.
4. Ibid., p. 88; Stride, P. 2008. *Journal of the Royal College of Physicians of Edinburgh* 38: 272–279.
5. Boswell, J. 1785. *The Journal of a Tour to the Hebrides with Samuel Johnson, LLD.* Henry Baldwin, London, pp. 220–221.
6. Maclean, C. 1977. *St Kilda: Island on the Edge of the World*, revised edition. Canongate, Edinburgh, p. 88.
7. Richards, E. 1992. *Scottish Historical Review* 71: 129–155.
8. Cook, N. D. 1998. *Born to Die: Disease and New World Conquest, 1492–1650.* Cambridge University Press, Cambridge.
9. Penman, B. S. *et al.* 2017. *Epidemiology & Infection* 145: 1–11.
10. Shanks, G. D. 2016. *American Journal of Tropical Medicine and Hygiene* 95: 273.
11. Ryan, P. 2000. *Fiji's Natural Heritage*, revised edition. Exisle Publishing, Auckland, p. 156.
12. Thompson, C. 2019. *Sea People.* William Collins, London, p. 154.
13. Bosma, U. 2015. *Bijdragen tot de Taal-, Land- en Volkenkunde* 171: 69–96.
14. Sacks, O. 1996. *The Island of the Colour-Blind.* Picador/Macmillan, London.
15. Bromham, L. & Cardillo, M. 2007. *Biology Letters* 3: 398–400.
16. Perry, G. H. & Dominy, N. J. 2009. *Trends in Ecology & Evolution* 24: 218–225.
17. Berger, L. R. *et al.* 2008. *PLoS ONE* 3: e1780.
18. Fitzpatrick, S. M. *et al.* 2008. *PLoS ONE* 3: e3015.
19. Latham, R. (transl. 1958). *The Travels of Marco Polo.* Penguin, London, p. 230.
20. Cipriani, L. (transl. 1966). *The Andaman Islanders.* Weidenfeld & Nicolson, London.
21. Dobson, G. E. 1875. *Journal of the Anthropological Institute of Great Britain and Ireland* 4: 459.

22. Thangaraj, K. *et al.* 2005. *Science* 308: 996; Aghakhanian, F. *et al.* 2015. *Genome Biology and Evolution* 7: 1206–1215; Jinam, T. A. *et al.* 2017. *Genome Biology and Evolution* 9: 2013–2022.

23. For further discussion see *Human Biology* 85: 7–20, 495–502.

24. Jinam, T. A. *et al.* 2017. *Genome Biology and Evolution* 9: 2013–2022; Mahdi, W. 2017. In *Spirits and Ships: Cultural Transfers in Early Monsoon Asia*. ISEAS–Yusof Ishak Institute, Singapore, pp. 325–440.

25. Diamond, J. 1997. *Guns, Germs, and Steel: A Short History of Everybody for the Last 13,000 Years*. Jonathan Cape, London.

26. Tucci, S. *et al.* 2018. *Science* 361: 511–516.

27. Gill, Rev. 1854. *Edinburgh New Philosophical Journal* 57: 144.

28. Hoskins, J. 1993. *The Play of Time: Kodi Perspectives on Calendars, History, and Exchange*. University of California Press, Berkeley and Los Angeles, CA, p. 80.

29. Matsu'ura, S. *et al.* 2020. *Science* 367: 210–214.

30. Augusta, J. & Burian, Z. 1960. *Prehistoric Man*. Paul Hamlyn, London, plate 22.

31. Rizal, Y. *et al.* 2019. *Nature* 577: 381–385.

32. Joordens, J. C. A. *et al.* 2015. *Nature* 518: 228–231.

33. O'Connor, S. *et al.* 2011. *Science* 334: 1117–1121.

34. Weidenreich, F. 1945. *Anthropological Papers of the American Museum of Natural History* 40: 5.

35. von Koenigswald, G. H. R. 1956. *Meeting Prehistoric Man*. Thames & Hudson, London, p. 111.

36. Weidenreich, F. 1946. *Apes, Giants, and Man*. University of Chicago Press, Chicago, IL, p. 48.

37. Ibid., p. 52.

38. Ibid., p. 61.

39. Zanolli, C. *et al.* 2019. *Nature Ecology & Evolution* 3: 755–764.

40. Maringer, J. & Verhoeven, T. 1979. *East and West* 29: 247–263.

41. Morwood, M. *et al.* 1998. *Nature* 392: 173–176; Morwood, M. *et al.* 1999. *Antiquity* 73: 273–286; Brumm, A. *et al.* 2006. *Nature* 441: 624–628.

42. Bednarik, R. G. 2003. *Cambridge Archaeological Journal* 13: 41–66.

43. von Koenigswald, G. H. R. 1956. *Meeting Prehistoric Man*. Thames & Hudson, London, p. 28.

44. Augusta, J. & Burian, Z. 1960. *Prehistoric Man*. Paul Hamlyn, London, plate 22.

45. Morwood, M. & van Oosterzee, P. 2007. *A New Human: The Startling Discovery and Strange Story of the 'Hobbits' of Flores, Indonesia*. Smithsonian Books/HarperCollins, New York, NY, pp. 36–38.

46. Brown, P. *et al.* 2004. *Nature* 431: 1055–1061.

47. Gee, H. 2004. Our not so distant relative, theguardian.com/education/2004/oct/28/research.highereducation1

48. Jacob, T. *et al.* 2006. *Proceedings of the National Academy of Sciences of the USA* 103: 13421–13426; Aiello, L. C. 2010. *American Journal of Physical Anthropology* 142: 167–179; Eckhardt, R. B. *et al.* 2014. *Proceedings of the National Academy of Sciences of the USA* 111: 11961–11966; Henneberg, M. *et al.* 2014. *Proceedings of the National Academy of Sciences of the USA* 111: 11967–11972.

49. Henneberg, M. & Schofield, J. 2008. *The Hobbit Trap: Money, Fame, Science and the Discovery of a 'New Species'*. Wakefield Press, Kent Town, SA; Jungers, W. L. & Kaifu, Y. 2011. *American Journal of Physical Anthropology* 145: 282–289.

50. Dalton, R. 2005. *Nature* 434: 5.

51. Morwood, M. J. *et al.* 2005. *Nature* 437: 1012–1017; van den Bergh, G. D. *et al.* 2016. *Nature* 534: 245–248.

52. Détroit, F. *et al.* 2019. *Nature* 568: 181–186.

53. Aubert, M. *et al.* 2019. *Nature* 576: 442–445.

54. Brumm, A. *et al.* 2021. *Science Advances* 7: eabd4648.

55. Pike, A. W. G. *et al.* 2012. *Science* 336: 1409–1413; Hoffmann, D. L. *et al.* 2018. *Science* 359: 912–915; Zhang, D. D. *et al.* 2021. *Science Bulletin* 66: 2506–2515.

56. Callaway, E. 2019. Is this cave painting humanity's oldest story?, nature.com/articles/d41586-019-03826-4

57. Sutikna, T. *et al.* 2016. *Nature* 532: 366–369.

58. The stories of the ebu gogo are from G. Forth 2008. *Images of the Wildman in Southeast Asia: An Anthropological Perspective*. Routledge, Abingdon.

59. Forth, G. 2012. In *The Anthropology of Extinction: Essays on Culture and Species Death*. Indiana University Press, Bloomington, IN, pp. 200–218.

60. Gee, H. 2004. Flores, God and Cryptozoology, nature.com/articles/news041025-2

61. Forth, G. 2022. *Between Ape and Human: An Anthropologist on the Trail of a Hidden Hominid*. Pegasus Books, New York, NY.

62. Forth, G. 1981. *Rindi: An Ethnographic Study of a Traditional Domain in Eastern Sumba*. Koninklijk Instituut voor Taal-, Land- en Volkenkunde, Leiden, v.

63. Weidenreich, F. 1946. *Apes, Giants, and Man*. University of Chicago Press, Chicago, IL, p. 66.

Chapter 4: Storytelling

1. Forth, G. 2019. *A Dog Pissing at the Edge of the Path: Animal Metaphors in an Eastern Indonesian Society*. McGill-Queen's University Press, Montreal and Kingston, p. 58.

2. Ibid., p. 218.

3. Burney, D. A. & Ramilisonina. 1998. *American Anthropologist* 100: 957–966.

4. Hoskins, J. 1993. *The Play of Time: Kodi Perspectives on Calendars, History, and Exchange.* University of California Press, Berkeley and Los Angeles, CA, p. 82.

5. Brookesmith, P. (ed.) 1984. *Creatures from Elsewhere: Weird Animals That No-One Can Explain.* Orbis, London.

6. Forth, G. 2008. *Images of the Wildman in Southeast Asia: An Anthropological Perspective.* Routledge, Abingdon.

7. Heuvelmans, B. 1995. *On the Track of Unknown Animals*, revised third English edition. Kegan Paul International, London, p. 181.

8. Coon, C. S. 1955. *The History of Man: From the First Human to Primitive Culture and Beyond.* Jonathan Cape, London, p. 28.

9. Heuvelmans, B. 1995. *On the Track of Unknown Animals*, revised third English edition. Kegan Paul International, London, p. 210.

10. Regal, B. 2009. *Annals of Science* 66: 83–102; Regal, B. 2013. *Searching for Sasquatch: Crackpots, Eggheads, and Cryptozoology.* Palgrave Macmillan, London.

11. Heuvelmans, B. (transl. 2016). *Neanderthal: The Strange Saga of the Minnesota Iceman.* Anomalist Books, San Antonio, TX.

12. Bayanov, D. & Bourtsev, I. 1976. *Current Anthropology* 17: 312.

13. Sykes, B. C. *et al.* 2014. *Proceedings of the Royal Society B* 281: 20140161; Sykes, B. 2015. *The Nature of the Beast.* Coronet, London.

14. Lan, T. *et al.* 2017. *Proceedings of the Royal Society B* 284: 20171804.

15. Shao, Q. *et al.* 2017. *Quaternary International* 434: 65–74.

16. Naish, D. 2017. *Hunting Monsters: Cryptozoology and the Reality Behind the Myths.* Sirius, London, p. 139.

17. Shah, H. 2018. The scientist Grover Krantz risked it all … chasing bigfoot, smithsonianmag.com/smithsonian-institution/scientist-grover-krantz-risked-it-all-chasing-bigfoot-180970676/

18. Dammerman, K. W. 1924. *De Tropische Natuur* 13: 177–182; Dammerman, K. W. 1929. *Proceedings of the 4th Pacific Science Congress* 3: 121–126; Dammerman, K. W. 1932. *De Tropische Natuur* 21: 123–131; Forth, G. 2008. *Images of the Wildman in Southeast Asia: An Anthropological Perspective.* Routledge, Abingdon.

19. MacKinnon, J. R. 1974. *In Search of the Red Ape.* Collins, London; Forth, G. 2008. *Images of the Wildman in Southeast Asia: An Anthropological Perspective.* Routledge, Abingdon; Naish, D. 2017. *Hunting Monsters: Cryptozoology and the Reality Behind the Myths.* Sirius, London, p. 219.

20. Osman Hill, W. C. 1945. *Loris* 4: 251–262.

21. Osman Hill, W. C. 1961. *Oryx* 6: 93.

22. Ibid., p. 97.

23. Ibid., p. 95.

24. Forth, G. 2012. *Anthropology Today* 28: 16.

Chapter 5: Rodents of Unusual Size

1. Hoskins, J. 1993. *The Play of Time: Kodi Perspectives on Calendars, History, and Exchange.* University of California Press, Berkeley and Los Angeles, CA, p. 2.
2. Eaton, J. A. *et al.* 2016. *Birds of the Indonesian Archipelago: Greater Sundas and Wallacea.* Lynx Edicions, Barcelona, p. 78; although see F. E. Rheindt *et al.* (2017; *Zootaxa* 4250: 401–433) for a different taxonomic interpretation.
3. Biknevicius, A. R. *et al.* 1993. *American Museum Novitates* 3079: 1–25.
4. Woods, R. *et al.* 2021. *Molecular Biology and Evolution* 38: 84–95.
5. McFarlane, D. A. *et al.* 1998. *Quaternary Research* 50: 80–89.
6. Rowe, K. C. *et al.* 2016. *Evolution* 70: 653–665.
7. Giant carnivorous rat with long pubic hair discovered in Indonesia, news.mongabay.com/2015/10/giant-carnivorous-rat-with-long-pubic-hair-discovered-in-indonesia/
8. Stümpke, H. 1967. *The Snouters: Form and Life of the Rhinogrades.* American Museum of Natural History/Natural History Press, New York, NY.
9. Esselstyn, J. A. *et al.* 2012. *Biology Letters* 8: 990–993.
10. Thomson, V. *et al.* 2014. *PLoS ONE* 9: e91356.
11. Caravaggi, A. *et al.* 2019. *Ibis* 161: 648–661.
12. Brace, S. *et al.* 2014. *Journal of Biogeography* 41: 1583–1589.
13. Barker, D. *et al.* 2005. *Notornis* 52: 143–149.
14. Miskelly, C. M. & Fraser, J. R. 2006. *Notornis* 53: 353–359.
15. Dransfield, J. *et al.* 1984. *Nature* 312: 750–752.
16. Hunt, T. & Lipo, C. 2011. *The Statues That Walked: Unraveling the Mystery of Easter Island.* Free Press, New York, NY, p. 2.
17. Ibid., p. 8.
18. Diamond, J. 2005. *Collapse: How Societies Choose to Fail or Succeed.* Viking, New York, NY, p. 118.
19. Ibid., p. 114.
20. Hunt, T. & Lipo, C. 2011. *The Statues That Walked: Unraveling the Mystery of Easter Island.* Free Press, New York, NY.
21. Athens, J. S. 2009. *Biological Invasions* 11: 1489–1501.
22. Graham, N. A. J. *et al.* 2018. *Nature* 559: 250–253.
23. Cooke, S. B. & Crowley, B. E. 2018. *Journal of Vertebrate Paleontology* 38: e1510414.
24. Wyatt, K. B. *et al.* 2008. *PLoS ONE* 3: e3602.
25. Miller, G. S. 1929. *Smithsonian Miscellaneous Collections* 82(5): 1–16.
26. Eaton, J. A. *et al.* 2016. *Birds of the Indonesian Archipelago: Greater Sundas and Wallacea.* Lynx Edicions, Barcelona, p. 190.
27. Riley, E. P. 2010. *Evolutionary Anthropology* 19: 22–36.
28. van den Bergh, G. D. *et al.* 2009. *Journal of Human Evolution* 57: 527–537.

29. Forth, G. 2016. *Why the Porcupine is Not a Bird: Explorations in the Folk Zoology of an Eastern Indonesian People.* University of Toronto Press, Toronto.

30. Cheke, A. & Hume, J. 2008. *Lost Land of the Dodo: An Ecological History of Mauritius, Réunion & Rodrigues.* T & AD Poyser, London.

31. Tyson, E. 1699. *Orang-Outang, sive Homo Sylvestris: Or, The Anatomy of a Pygmie Compared with that of a Monkey, an Ape, and a Man. To which is Added, a Philological Essay Concerning the Pygmies, the Cynocephali, the Satyrs, and Sphinges of the Ancients.* T. Bennett & D. Brown, London.

32. Sunderland-Groves, J. L. 1990. *Population, Distribution and Conservation Status of the Cross River Gorilla (Gorilla gorilla diehli) in Cameroon.* Master's thesis, University of Sussex; Lingomo, B. & Kimura, D. 2009. *African Study Monographs* 30: 209–225; Zhang, L. *et al.* 2020. *Biological Conservation* 241: 108267.

33. Skott, C. 2014. *Indonesia and the Malay World* 42: 150.

34. Corbey, R. 2005. *The Metaphysics of Apes: Negotiating the Animal–Human Boundary.* Cambridge University Press, Cambridge; Blancke, S. 2014. In *The Evolution of Social Communication in Primates.* Springer, New York, NY, pp. 31–44.

35. Skott, C. 2014. *Indonesia and the Malay World* 42: 159.

36. Lovejoy, A. O. 1904. *Popular Science Monthly* 65: 340; Sebastiani, S. 2022. *History of European Ideas* 48: 51.

37. Blancke, S. 2014. In *The Evolution of Social Communication in Primates.* Springer, New York, NY, p. 33.

38. Skott, C. 2014. *Indonesia and the Malay World* 42: 148.

39. Ibid., p. 151.

40. Ibid., p. 150.

41. Ibid., pp. 145–146.

42. Leatherdale, D. 2017. Was a monkey really hanged in Hartlepool?, bbc.co.uk/news/uk-england-tees-40801937

43. Forth, G. 2008. *Images of the Wildman in Southeast Asia: An Anthropological Perspective.* Routledge, Abingdon, p. 132.

44. Dammerman, K. W. 1924. *De Tropische Natuur* 13: 177–182; Dammerman, K. W. 1929. *Proceedings of the 4th Pacific Science Congress* 3: 121–126; Dammerman, K. W. 1932. *De Tropische Natuur* 21: 123–131; Forth, G. 2008. *Images of the Wildman in Southeast Asia: An Anthropological Perspective.* Routledge, Abingdon; Meijaard, E. *et al.* 2021. *PLoS ONE* 16: e0238087.

45. Stonor, C. 1955. *The Sherpa and the Snowman.* Hollis & Carter, London, p. 30.

46. Forth, G. 2008. *Images of the Wildman in Southeast Asia: An Anthropological Perspective.* Routledge, Abingdon, pp. 101–102; Forth, G. 2022. *Between Ape and Human: An Anthropologist on the Trail of a Hidden Hominid.* Pegasus Books, New York, NY, pp. 188–190.

47. Centlivres, P. & Girod, I. 1998. *Gradhiva* 24: 33–43.

48. Dammerman, K. W. 1932. *De Tropische Natuur* 21: 123–131.
49. Messner, R. 1998 (transl. 2000). *My Quest for the Yeti: Confronting the Himalayas' Deepest Mystery.* Macmillan, London.
50. Latham, R. (transl. 1958). *The Travels of Marco Polo.* Penguin, London, p. 226.
51. Forth, G. 2008. *Images of the Wildman in Southeast Asia: An Anthropological Perspective.* Routledge, Abingdon, p. 83.
52. Forth, G. 2012. *Anthropology Today* 28: 16; Forth, G. 2022. *Between Ape and Human: An Anthropologist on the Trail of a Hidden Hominid.* Pegasus Books, New York, NY, pp. 77–78.

Chapter 6: Tulang Junkie

1. Rudwick, M. J. S. 1997. *Georges Cuvier, Fossil Bones, and Geological Catastrophes.* University of Chicago Press, Chicago, IL.
2. Lovejoy, A. O. 1936. *The Great Chain of Being.* Harvard University Press, Cambridge, MA.
3. Rudwick, M. J. S. 1976. *The Meaning of Fossils: Episodes in the History of Palaeontology.* University of Chicago Press, Chicago, IL.
4. Jefferson, T. 1788. *Notes on the State of Virginia.* Prichard and Hall, Philadelphia, PA, p. 54.
5. For extensive background on early interpretations of fossils, see: Rudwick, M. J. S. 1976. *The Meaning of Fossils: Episodes in the History of Palaeontology.* University of Chicago Press, Chicago, IL; Cutler, A. 2004. *The Seashell on the Mountaintop.* Arrow Books, London; Jordan, J. M. 2016. *History and Philosophy of the Life Sciences* 38: 90–116.
6. Hellyer, M. 1996. *Archives of Natural History* 23: 43–60.
7. Fortey, R. 2008. *Nature* 455: 35.
8. Cutler, A. 2004. *The Seashell on the Mountaintop.* Arrow Books, London, p. 195.
9. Woodward, J. 1695. *An Essay Toward a Natural History of the Earth: And Terrestrial Bodies, Especially Minerals; As also of the Sea, Rivers, and Springs. With an Account of the Universal Deluge: And of the Effects that it had upon the Earth.* Richard Wilkin, London, pp. A5–A6.
10. Ibid., p. A5.
11. Wendt, H. (transl. 1968). *Before the Deluge.* Victor Gollancz, London, pp. 44–45.
12. Stone, R. 2002. *Mammoth: The Resurrection of an Ice Age Giant.* 4th Estate, London, p. 22.
13. McKay, J. J. 2017. *Discovering the Mammoth: A Tale of Giants, Unicorns, Ivory, and the Birth of a New Science.* Pegasus Books, New York, NY, p. 91.
14. Weishampel, D. B. & White, N. M. (eds.) 2003. *The Dinosaur Papers: 1676–1906.* Smithsonian Books, Washington, DC, p. 12.

15. McKay, J. J. 2017. *Discovering the Mammoth: A Tale of Giants, Unicorns, Ivory, and the Birth of a New Science.* Pegasus Books, New York, NY, pp. 164–168.

16. Jefferson, T. 1788. *Notes on the State of Virginia.* Prichard and Hall, Philadelphia, PA, p. 54.

17. Gray, H. G. & Cotton, M. A. 1966. *The Meare Lake Village: A Full Description of the Excavations and Relics from the Eastern Half of the West Village, 1910–1933.* Volume III. Taunton Castle, Taunton, p. 410; Coles, J. & Minnitt, S. 1995. *Industrious and Fairly Civilized: The Glastonbury Lake Village.* Somerset Levels Project/Somerset County Council Museums Service, Taunton, p. 65.

18. Secher, A. 2022. *Travels with Trilobites: Adventures in the Paleozoic.* Columbia University Press, New York, NY, p. 345.

19. Mayor, A. 2000. *The First Fossil Hunters: Paleontology in Greek and Roman Times.* Princeton University Press, Princeton, NJ and Oxford, p. 71.

20. Cutler, A. 2004. *The Seashell on the Mountaintop,* Arrow Books, London, p. 56.

21. Crump, J. & Crump, I. 1963. *Dragon Bones in the Yellow Earth.* Dodd, Mead & Company, New York, NY, p. 18.

22. North, F. J. 1964. *The Evolution of the Bristol Channel,* third edition. National Museum of Wales, Cardiff, p. 60.

23. Williams, M. 2021. *When the Sahara Was Green.* Princeton University Press, Princeton, NJ, pp. 23, 33–34.

24. Wilford, J. N. 1994. World's oldest paved road found in Egypt, nytimes.com/1994/05/08/world/world-s-oldest-paved-road-found-in-egypt.html

25. Gallenkamp, C. 2001. *Dragon Hunter: Roy Chapman Andrews and the Central Asiatic Expeditions.* Penguin, New York, NY, p. 245.

26. Owens, R. M. 1984. *Trilobites in Wales.* National Museum of Wales, Cardiff, p. 5.

27. Evans, G. E. 2013. *The Pattern Under The Plough: Aspects of the Folk Life of East Anglia.* Little Toller Books, Toller Fratrum, p. 123.

28. Johnson, C. 2006. *Australia's Mammal Extinctions: A 50 000 Year History.* Cambridge University Press, Melbourne, p. 71.

29. Shuker, K. P. N. 2013. *Mirabilis: A Carnival of Cryptozoology and Unnatural History.* Anomalist Books, San Antonio, TX, p. 64.

30. Haddon, A. C. 1932. *Head-Hunters Black, White, and Brown.* Watts & Co, London, p. 204.

31. Merrifield, R. 1987. *The Archaeology of Ritual and Magic.* B.T. Batsford, London, p. 16; Evans, G. E. 2013. *The Pattern Under The Plough: Aspects of the Folk Life of East Anglia.* Little Toller Books, Toller Fratrum, p. 126.

32. von Koenigswald, G. H. R. 1956. *Meeting Prehistoric Man.* Thames & Hudson, London, p. 134.

33. Duffin, C. J. 2008. *Ferrantia* 54: 13.
34. Ibid., p. 15.
35. Ibid., p. 16.
36. Mayor, A. 2000. *The First Fossil Hunters: Paleontology in Greek and Roman Times*. Princeton University Press, Princeton, NJ and Oxford, pp. 55–60.
37. Jochelson, W. 1909. *American Naturalist* 43: 48.
38. Berkes, F. 2018. *Sacred Ecology*, fourth edition. Routledge, New York, NY and Abingdon, p. 113.
39. Romano, M. & Avanzini, M. 2019. *Historical Biology* 31: 122.
40. Flannery, T. 2018. *Europe: A Natural History*. Allen Lane, London, p. 246.
41. Mayor, A. 2000. *The First Fossil Hunters: Paleontology in Greek and Roman Times*. Princeton University Press, Princeton, NJ and Oxford, p. 71. Although for an alternative opinion, see: markwitton-com.blogspot.co .uk/2016/04/why-protoceratops-almost-certainly.html
42. Stone, R. 2002. *Mammoth: The Resurrection of an Ice Age Giant*. 4th Estate, London, p. 22.
43. Uchiyama, J. 2020. *Auspicious Animals: The Art of Good Omens*. PIE International, Tokyo, p. 57.
44. Rosendahl, W. *et al.* 2005. *Abhandlungen der Naturhistorischen Gesellschaft zu Nürnberg* 45: 199–213; McKay, J. J. 2017. *Discovering the Mammoth: A Tale of Giants, Unicorns, Ivory, and the Birth of a New Science*. Pegasus Books, New York, NY.
45. McKay, J. J. 2017. *Discovering the Mammoth: A Tale of Giants, Unicorns, Ivory, and the Birth of a New Science*. Pegasus Books, New York, NY; Romano, M. & Avanzini, M. 2019. *Historical Biology* 31: 117–139.
46. van der Geer, A. *et al.* 2008. *Folklore* 119: 71–92.
47. Romano, M. & Avanzini, M. 2019. *Historical Biology* 31: 121–122.
48. Ibid., p. 131.
49. Ibid., p. 122.
50. Shindler, K. 2005. *Discovering Dorothea: The Life of the Pioneering Fossil-Hunter Dorothea Bate*. HarperCollins, London; van der Geer, A. *et al.* 2010. *Evolution of Island Mammals: Adaptation and Extinction of Placental Mammals on Islands*. Wiley-Blackwell, Chichester, pp. 35–37.
51. Colbert, E. H. 1971. *Men and Dinosaurs*. Pelican, Harmondsworth, p. 23.
52. White, A. D. 1896. *A History of the Warfare of Science with Theology in Christendom*, vol. 2. D. Appleton & Company, New York, NY, p. 29.
53. van der Geer, A. *et al.* 2008. *Folklore* 119: 72.
54. Gibbons, A. 2019. *Science* 364: 418–419.
55. Romano, M. & Avanzini, M. 2019. *Historical Biology* 31: 130.
56. Mayor, A. 2000. *The First Fossil Hunters: Paleontology in Greek and Roman Times*. Princeton University Press, Princeton, NJ and Oxford, pp. 77, 293.

57. Romano, M. & Avanzini, M. 2019. *Historical Biology* 31: 125.
58. Fagan, B. 2018. *A Little History of Archaeology.* Yale University Press, New Haven, CT, p. 62.
59. Romano, M. & Avanzini, M. 2019. *Historical Biology* 31: 135.
60. Weishampel, D. B. & White, N. M. (eds.) 2003. *The Dinosaur Papers: 1676–1906.* Smithsonian Books, Washington, DC, p. 12.
61. Forth, G. 2016. *Why the Porcupine is Not a Bird: Explorations in the Folk Zoology of an Eastern Indonesian People.* University of Toronto Press, Toronto, pp. 311–312.
62. Forth, G. 2006. *Bijdragen tot de Taal-, Land- en Volkenkunde* 162: 336–349; Forth, G. 2022. *Between Ape and Human: An Anthropologist on the Trail of a Hidden Hominid.* Pegasus Books, New York, NY.
63. Forth, G. 1981. *Rindi: An Ethnographic Study of a Traditional Domain in Eastern Sumba.* Koninklijk Instituut voor Taal-, Land- en Volkenkunde, Leiden, p. 176.

Chapter 7: The Wall of the Mili Mongga

1. Malson, L. (transl. 1972). *Wolf Children.* NLB, London, p. 44.
2. Favazza, A. R. 1977. *British Journal of Medical Psychology* 50: 105. For footnote, see R. Nash, 2003. *Wild Enlightenment: The Borders of Human Identity in the Eighteenth Century.* University of Virginia Press, Charlottesville, VA and London.
3. Lane, M. 2011. Who was Peter the Wild Boy? bbc.co.uk/news/magazine -14215171
4. Malson, L. (transl. 1972). *Wolf Children.* NLB, London, pp. 9–10.
5. Sykes, B. 2015. *The Nature of the Beast.* Coronet, London, pp. 23–26.
6. Frankcom, G. & Musgrave, J. H. 1976. *The Irish Giant.* Gerald Duckworth, London, p. 15.
7. Forth, G. 2008. *Images of the Wildman in Southeast Asia: An Anthropological Perspective.* Routledge, Abingdon, p. 105.
8. Messner, R. (transl. 2000). *My Quest for the Yeti: Confronting the Himalayas' Deepest Mystery.* Macmillan, London, p. 78.
9. Skott, C. 2014. *Indonesia and the Malay World* 42: 157.
10. Ibid., p. 155.
11. Ibid., pp. 159–160.
12. Sterndale, R. A. 1884. *Natural History of the Mammalia of India and Ceylon.* Thacker, Spink, and Co, London, p. 6.
13. de Sepulveda, J. G. 1892. *Boletín de la Real Academia de la Historia,* vol. XXI.
14. Strassberg, R. E. 2002. *A Chinese Bestiary: Strange Creatures from the Guideways Through Mountains and Seas.* University of California Press, Berkeley and Los Angeles, CA.

15. Nappi, C. 2009. *The Monkey and the Inkpot: Natural History and Its Transformations in Early Modern China*. Harvard University Press, Cambridge, MA, pp. 129–130.
16. Dikötter, F. 2015. *The Discourse of Race in Modern China*, second edition. Oxford University Press, Oxford, p. 3.
17. Ibid.
18. Smith, O. D. 2021. *Sino-Platonic Papers* 309: 1–17.
19. Thompson, C. 2019. *Sea People*. William Collins, London, pp. 5–7.
20. Connolly, B. & Anderson, R. 1987. *First Contact: New Guinea's Highlanders Encounter the Outside World*. Viking, New York, NY, p. 47.
21. Ibid., p. 41.
22. Ibid., pp. 43–44.
23. Ibid., p. 46.
24. Bubandt, N. 2019. *Journal of the Royal Anthropological Institute* 25: 223–240.
25. MacCulloch, J. A. 1932. *Folklore* 43: 362.
26. Rees, A. 2019. *History of the Human Sciences* 32: 105.
27. Silver, C. 1986. *Browning Institute Studies* 14: 141–156.
28. Ibid.; MacCulloch, J. A. 1932. *Folklore* 43: 362–375.
29. Machen, A. 2007. *Dreads and Drolls*. Tartarus Press, Leyburn, p. 215.
30. MacCulloch, J. A. 1932. *Folklore* 43: 366.
31. Sykes, B. 2015. *The Nature of the Beast*. Coronet, London.
32. Carlhoff, S. *et al.* 2021. *Nature* 596: 543–547.
33. Maloney, T. R. *et al.* 2022. *Nature* 609: 547–551.
34. Thompson, C. 2019. *Sea People*. William Collins, London, pp. 8–9.
35. Lansing, J. S. *et al.* 2007. *Proceedings of the National Academy of Sciences of the USA* 104: 16025.
36. Ibid.
37. Tumonggor, M. K. *et al.* 2014. *Journal of Human Genetics* 59: 494–503; Gomes, S. M. *et al.* 2017. *European Journal of Human Genetics* 25: 246–252.
38. Lansing, J. S. *et al.* 2007. *Proceedings of the National Academy of Sciences of the USA* 104: 16025.
39. Tucci, S. *et al.* 2018. *Science* 361: 511–516.
40. Skott, C. 2014. *Indonesia and the Malay World* 42: 160.
41. Hamilton, A. 2006. *Journal of Southeast Asian Studies* 37: 306.
42. Evans, I. H. N. 1925. *Journal of the Federated Malay States Museums* 12: 34–58; Brandt, J. H. 1961. *Journal of the Siam Society* 49: 124–125.
43. Forth, G. 2008. *Images of the Wildman in Southeast Asia: An Anthropological Perspective*. Routledge, Abingdon, p. 108.
44. Hung, H. C. *et al.* 2022. *World Archaeology* 54: 207–228.
45. Tregear, E. 1888. *Journal of the Anthropological Institute of Great Britain and Ireland* 17: 303.
46. Krappe, A. H. 1930. *The Science of Folk-Lore*. Methuen & Co, London, p. 87.

47. Hoskins, J. 1993. *The Play of Time: Kodi Perspectives on Calendars, History, and Exchange*. University of California Press, Berkeley and Los Angeles, CA, p. 32.

48. Forth, G. 2008. *Images of the Wildman in Southeast Asia: An Anthropological Perspective*. Routledge, Abingdon, p. 53.

49. arkhamarchivist.com/wordcount-lovecraft-favorite-words/

50. Romano, M. & Avanzini, M. 2019. *Historical Biology* 31: 130.

51. Hulse, M. 1992. Introduction to *Caspar Hauser: The Inertia of the Heart*. Penguin, London, p. xi.

52. Krappe, A. H. 1930. *The Science of Folk-Lore*. Methuen & Co, London. p. 75.

53. Hoskins, J. 1993. *The Play of Time: Kodi Perspectives on Calendars, History, and Exchange*. University of California Press, Berkeley and Los Angeles, CA, p. 34.

54. Hoskins, J. 1996. In *Headhunting and the Social Imagination in Southeast Asia*. Stanford University Press, Stanford, CA, p. 221.

55. Hoskins, J. 1993. *The Play of Time: Kodi Perspectives on Calendars, History, and Exchange*. University of California Press, Berkeley and Los Angeles, CA, p. 48.

56. Ibid.

57. Bosma, U. 2015. *Bijdragen tot de Taal-, Land- en Volkenkunde* 171: 69–96.

58. Needham, R. 1983. *Sumba and the Slave Trade*. Centre of Southeast Asian Studies, Monash University, Melbourne.

59. Ibid., p. 39.

60. Ibid., p. 15.

61. Hoskins, J. 1996. In *Headhunting and the Social Imagination in Southeast Asia*. Stanford University Press, Stanford, CA, p. 222.

62. Needham, R. 1983. *Sumba and the Slave Trade*. Centre of Southeast Asian Studies, Monash University, Melbourne, p. 41.

63. Ibid.

64. Hoskins, J. 1996. In *Headhunting and the Social Imagination in Southeast Asia*. Stanford University Press, Stanford, CA, p. 226.

65. Turvey, S. T. *et al.* 2018. *Science* 360: 1346–1349.

66. Malek, J. 1993. *The Cat in Ancient Egypt*. British Museum Press, London; 70 million animal mummies: Egypt's dark secret, bbc.co.uk/news/scien ce-environment-32685945

67. van der Grijp, P. 2007. *Journal of the Polynesian Society* 116: 353.

68. Adams, R. 2007. *The Megalithic Tradition of West Sumba, Indonesia: An Ethnoarchaeological Investigation of Megalith Construction*. PhD thesis, Simon Fraser University, p. 137.

69. Hoskins, J. 1996. In *Headhunting and the Social Imagination in Southeast Asia*. Stanford University Press, Stanford, CA, pp. 216–248.

70. Hoskins, J. 1993. *The Play of Time: Kodi Perspectives on Calendars, History, and Exchange*. University of California Press, Berkeley and Los Angeles, CA, p. 312.

Chapter 8: An Interlude with Giant Rats

1. Turvey, S. T. *et al.* 2017. *Proceedings of the Royal Society B* 284: 20171278.
2. Veatch, E. G. *et al.* 2019. *Journal of Human Evolution* 130: 45–60.
3. Forth, G. 2016. *Why the Porcupine is Not a Bird: Explorations in the Folk Zoology of an Eastern Indonesian People.* University of Toronto Press, Toronto, p. 117.
4. Turvey, S. T. *et al.* 2012. *Mammalian Biology* 77: 404–413; Mistretta, B. A. *et al.* 2021. *Zootaxa* 4951: 434–460.
5. Forth, G. 2016. *Why the Porcupine is Not a Bird: Explorations in the Folk Zoology of an Eastern Indonesian People.* University of Toronto Press, Toronto, pp. 120–122.
6. Flannery, T. 1998. *Throwim Way Leg: Adventures in the Jungles of New Guinea.* Phoenix, London, p. 238.
7. Coles, B. 2019. *Avanke, Bever, Castor: The Story of Beavers in Wales.* Wetland Archaeology Research Project/Hedgerow Print, Thorverton, p. 36.
8. Belarus: Man dies after being attacked by beaver, bbc.co.uk/news/av/world-europe-22707094
9. Wendt, H. (transl. 1968). *Before the Deluge.* Victor Gollancz, London, p. 34.
10. Burnham, K. K. *et al.* 2009. *Ibis* 151: 514–522.
11. Cheng, H. *et al.* 2020. *Proceedings of the National Academy of Sciences of the USA* 117: 23408–23417.
12. Alroy, J. 2001. *Science* 292: 1893–1896; Johnson, C. 2006. *Australia's Mammal Extinctions: A 50 000 Year History.* Cambridge University Press, Melbourne, pp. 106–111.
13. Grayson, D. K. 2008. *Proceedings of the National Academy of Sciences of the USA* 105: 4077–4078.
14. Clement, C. R. 2015. *Proceedings of the Royal Society B* 282: 20150813.
15. Ruddiman, W. F. 2005. *Plows, Plagues, and Petroleum: How Humans Took Control of Climate.* Princeton University Press, Princeton, NJ.
16. Turvey, S. T. (ed.) 2009. *Holocene Extinctions.* Oxford University Press, Oxford.
17. Johnson, C. N. & Wroe, S. 2003. *Holocene* 13: 941–948.
18. Fowler, C. 2013. *Ignition Stories: Indigenous Fire Ecology in the Indo-Australian Monsoon Zone.* Carolina Academic Press, Durham, NC, p. 225.
19. Monk, K. A. *et al.* 1997. *The Ecology of Nusa Tenggara and Maluku.* Oxford University Press, Oxford, p. 505.
20. Fowler, C. 2013. *Ignition Stories: Indigenous Fire Ecology in the Indo-Australian Monsoon Zone.* Carolina Academic Press, Durham, NC, p. 10.
21. Johnson, C. 2006. *Australia's Mammal Extinctions: A 50 000 Year History.* Cambridge University Press, Melbourne.
22. Monk, K.A. *et al.* 1997. *The Ecology of Nusa Tenggara and Maluku.* Oxford University Press, Oxford, p. 288.

23. Fowler, C. 2013. *Ignition Stories: Indigenous Fire Ecology in the Indo-Australian Monsoon Zone.* Carolina Academic Press, Durham, NC, p. 30.

24. Monk, K. A. *et al.* 1997. *The Ecology of Nusa Tenggara and Maluku.* Oxford University Press, Oxford, p. 287.

25. Fowler, C. 2013. *Ignition Stories: Indigenous Fire Ecology in the Indo-Australian Monsoon Zone.* Carolina Academic Press, Durham, NC, pp. 115–116.

26. Forth, G. 2016. *Why the Porcupine is Not a Bird: Explorations in the Folk Zoology of an Eastern Indonesian People.* University of Toronto Press, Toronto, p. 119.

27. Fowler, C. 2013. *Ignition Stories: Indigenous Fire Ecology in the Indo-Australian Monsoon Zone.* Carolina Academic Press, Durham, NC, p. 116.

28. Gunn, G. C. 2016. *Review of Culture* 53: 125–148.

29. Turvey, S. T. *et al.* 2017. *Proceedings of the Royal Society B* 284: 20171278.

Chapter 9: The Island of the Day Before

1. Godbey, A. H. 1939. *American Journal of Semitic Languages and Literatures* 56: 269.

2. Dammerman, K. W. 1928. *Treubia* 10: 315.

3. Verhoeven, T. 1959. *Anthropos* 54: 970–972.

4. Hoskins, J. 1998. *Biographical Objects: How Things Tell the Stories of People's Lives.* Routledge, New York, NY and London, p. 119.

5. Liu, S. 2015. *Orientations* 46: 60–61.

6. Eliot, H. 2018. *The Penguin Classics Book.* Penguin, London, p. 107.

7. Pfennigwerth, S. 2010. *New Creatures Made Known: (Re)discovering the King Island Emu.* Master's thesis, University of Tasmania; Pfennigwerth, S. 2010. *Archives of Natural History* 37: 74–90.

8. Kumar, A. B. *et al.* 2017. *Journal of Crustacean Biology* 37: 157–167.

9. Forth, G. 2017. *Herpetological Review* 48: 304–310; Forth, G. 2019. *Crustaceana* 92: 921–941.

10. Clunie, F. 1999. *Birds of the Fiji Bush,* second edition. Fiji Museum, Suva, p. 123.

11. Wolfe, R. 2003. *Moa: The Dramatic Story of the Discovery of a Giant Bird.* Penguin, Auckland, p. 177.

12. Mayr, E. 1963. *Animal Species and Evolution.* Belknap Press, Cambridge, MA, p. 17.

13. Bauer, A. M. & Russell, A. P. 1987. *New Zealand Journal of Zoology* 13: 141–148.

14. Godfrey, L. R. & Jungers, W. L. 2003. *Evolutionary Anthropology* 12: 252–263.

15. Teit, J. A. 1917. *Journal of American Folklore* 30: 450–451.

16. Beck, J. C. 1972. *Ethnohistory* 19: 120.

17. Mayor, A. 2000. *The First Fossil Hunters: Paleontology in Greek and Roman Times.* Princeton University Press, Princeton, NJ and Oxford, p. 319.

18. Johnson, C. 2006. *Australia's Mammal Extinctions: A 50 000 Year History.* Cambridge University Press, Melbourne, pp. 70–71.

19. Nunn, P. 2018. *The Edge of Memory: Ancient Stories, Oral Tradition and the Post-Glacial World.* Bloomsbury Sigma, London, pp. 198, 266.

20. Wehi, P. M. *et al.* 2018. *Human Ecology* 46: 465.

21. Tregear, E. 1888. *Journal of the Anthropological Institute of Great Britain and Ireland* 17: 294.

22. Wehi, P. M. *et al.* 2018. *Human Ecology* 46: 461–470.

23. Tregear, E. 1904. *The Maori Race.* Archibald Dudingston Willis, Wanganui, p. 185.

24. Tregear, E. 1888. *Journal of the Anthropological Institute of Great Britain and Ireland* 17: 305.

25. Kanagavel, A. *et al.* 2017. *Ambio* 46: 695–705.

26. Burney, D. A. & Ramilisonina. 1998. *American Anthropologist* 100: 957–966.

27. For example, see R. Mabey, 1995. *The Oxford Book of Nature Writing.* Oxford University Press, Oxford, p. 13.

28. Borges, J. L. (transl. 2000). *The Total Library: Non-Fiction 1922–1986.* Allen Lane, London, p. 231.

29. Liu, S. 2015. *Orientations* 46: 61.

30. Barber, A. D. 2015. *Bulletin of the British Myriapod and Isopod Group* 28: 54–63; Hitchings, C. 2017. Map reveals surprising number of names for a very common garden species, bbc.co.uk/blogs/natureuk/entries/8d e8208d-cc92-4120-872b-340f05380e69

31. Fleck, D. W. *et al.* 1999. *International Journal of Primatology* 20: 1005–1028.

32. Voss, R. S. *et al.* 2014. *Journal of Mammalogy* 95: 893–898.

33. Leroi, A. M. 2014. *The Lagoon: How Aristotle Invented Science.* Bloomsbury, London, pp. 46, 104.

34. Fowler, C. S. & Turner, N. J. 1999. In *The Cambridge Encyclopedia of Hunters and Gatherers.* Cambridge University Press, Cambridge, pp. 419–420.

35. Lévi-Strauss, C. (transl. 1963). *Totemism.* Beacon Press, Boston.

36. Sterckx, R. 2019. *Chinese Thought: From Confucius to Cook Ding.* Pelican, London, p. 344.

37. Woods, C. A. & Ottenwalder, J. A. 1992. *The Natural History of Southern Haiti.* Florida Museum of Natural History, Gainesville, p. 112.

38. Turvey, S. T. *et al.* 2010. *Conservation Biology* 24: 778–787.

39. Tregear, E. 1888. *Journal of the Anthropological Institute of Great Britain and Ireland* 17: 305.

40. Coles, B. 2019. *Avanke, Bever, Castor: The Story of Beavers in Wales.* Wetland Archaeology Research Project/Hedgerow Print, Thorverton, p. 43.

41. Forth, G. 2021. *Anthrozoös* 34: 61–76.

42. Cheke, A. S. & Parish, J. C. 2020. *Quaternary* 3: 4.
43. Turvey, S. T. *et al.* 2019. *Philosophical Transactions of the Royal Society B* 374: 20190217.
44. Turvey, S. T. *et al.* 2018. *Royal Society Open Science* 5: 172352.
45. Buick, T. L. 1931. *The Mystery of the Moa*. Thomas Avery & Sons, New Plymouth.
46. Evans, E. E. 1970. *Where Beards Wag All: The Relevance of the Oral Tradition*. Faber & Faber, London, p. 224.
47. Ibid., p. 229.
48. Henige, D. 2009. *History in Africa* 36: 127.
49. Raglan, F. R. S. 2003. *The Hero: A Study in Tradition, Myth and Drama*. Dover, Mineola, NY, p. 3.
50. Ibid., p. 6.
51. Ibid., p. 9.
52. Ibid., p. 14.
53. Buckingham, W. 2018. *Stealing with the Eyes: Imaginings and Incantations in Indonesia*. Haus Publishing, London, p. 60.
54. Ditlevsen, T. (transl. 1985). *Childhood*. Penguin, London, p. 72.
55. Everett, D. 2008. *Don't Sleep, There Are Snakes: Life and Language in the Amazonian Jungle*. Profile Books, London.
56. Thorpe, L. (transl. 1966). *The History of the Kings of Britain*. Penguin, London, p. 19.
57. Machen, A. 2018. *The Great God Pan and Other Horror Stories*. Oxford University Press, Oxford, p. 135.
58. Spence, L. 1921. *Introduction to Mythology*. George G. Harrap, London.
59. Turner, R. C. 1986. In *Lindow Man: The Body in the Bog*. Guild Publishing, London, pp. 171–172.
60. Hodgen, M. T. 1931. *American Anthropologist* 33: 307.
61. Graça da Silva, S. & Tehrani, J. J. 2016. *Royal Society Open Science* 3: 150645; Pagel, M. 2016. *Current Biology* 26: R279–R281.
62. Henige, D. 2009. *History in Africa* 36: 132–133.
63. Vansina, J. 1985. *Oral Tradition as History*. James Currey, Woodbridge and Rochester, NY.
64. Thompson, C. 2019. *Sea People*. William Collins, London, p. 158.
65. Ibid., p. 128.
66. Hoskins, J. 1993. *The Play of Time: Kodi Perspectives on Calendars, History, and Exchange*. University of California Press, Berkeley and Los Angeles, CA, p. 17.
67. Nunn, P. 2018. *The Edge of Memory: Ancient Stories, Oral Tradition and the Post-Glacial World*. Bloomsbury Sigma, London, pp. 68, 229.
68. Kirch, P. V. 2017. *On the Road of the Winds*, revised edition. University of California Press, Oakland, CA.
69. Parker Pearson, M. *et al.* 2021. *Antiquity* 95: 85–103.

70. Veski, S. *et al.* 2001. *Meteoritics & Planetary Science* 36: 1367–1375.

71. Nunn, P. 2018. *The Edge of Memory: Ancient Stories, Oral Tradition and the Post-Glacial World.* Bloomsbury Sigma, London, p. 12.

72. Nunn, P. D. & Reid, M. J. 2016. *Australian Geographer* 47: 37.

73. Wilkie, B. *et al.* 2020. *Journal of Volcanology and Geothermal Research* 403: 106999.

74. Henige, D. 2009. *History in Africa* 36: 167.

75. Hamacher, D. W. & Goldsmith, J. 2013. *Journal of Astronomical History and Heritage* 16: 295–311.

76. Wilkie, B. *et al.* 2020. *Journal of Volcanology and Geothermal Research* 403: 106999.

77. Tankersley, K. B. *et al.* 2022. *Scientific Reports* 12: 1706; Neuhäuser, R. & Neuhäuser, D. L. 2022. *Scientific Reports* 12: 12090.

78. Henige, D. 2009. *History in Africa* 36: 184–185.

79. Hoskins, J. 1993. *The Play of Time: Kodi Perspectives on Calendars, History, and Exchange.* University of California Press, Berkeley and Los Angeles, CA, p. 44.

80. Lansing, J. S. *et al.* 2007. *Proceedings of the National Academy of Sciences of the USA* 104: 16022–16026.

81. Hoskins, J. 1993. *The Play of Time: Kodi Perspectives on Calendars, History, and Exchange.* University of California Press, Berkeley and Los Angeles, CA, pp. 99, 350.

82. Fowler, C. 2013. *Ignition Stories: Indigenous Fire Ecology in the Indo-Australian Monsoon Zone.* Carolina Academic Press, Durham, NC, p. 16.

83. Ibid., p. 126.

84. Hoskins, J. 1998. *Biographical Objects: How Things Tell the Stories of People's Lives.* Routledge, New York, NY and London, p. 150.

85. van Steenis, C. J. J. G. 1947. *Chronica Naturae* 103: 237–239.

86. Aubrey, J. 1862. *Wiltshire: The Topographical Collections of John Aubrey, F.R.S., A.D. 1659–70, with Illustrations. Corrected and Enlarged by John Edward Jackson, M.A., F.S.A.* Wiltshire Archaeological and Natural History Society, Devizes, p. 17.

Chapter 10: They Might Be Giants

1. Grinsell, L. V. 1976. *Folklore of Prehistoric Sites in Britain.* David & Charles, Newton Abbot, pp. 29, 92, 196.

2. Colquhoun, I. 1957. *The Living Stones: Cornwall.* Peter Owen, London and Chicago, IL, p. 212.

3. Grinsell, L. V. 1936. *The Ancient Burial-Mounds of England.* Methuen & Co, London, p. 41.

4. Gilchrist, R. & Green, C. 2015. *Glastonbury Abbey: Archaeological Excavations 1904–79.* Society of Antiquaries of London, London, p. 60.

5. Minozzi, S. *et al.* 2012. *Journal of Clinical Endocrinology & Metabolism* 97: 4302–4303.

6. Turner, R. C. 1986. In *Lindow Man: The Body in the Bog.* Guild Publishing, London, pp. 170–176.

7. Wood, E. J. 1868. *Giants and Dwarfs.* Richard Bentley, London, p. 2.

8. Ibid., p. 4.

9. Grafton, A. 2001. *Bring Out Your Dead: The Past as Revelation.* Harvard University Press, Cambridge, MA, p. 276.

10. Vico, G. (transl. 1984). *The New Science of Giambattista Vico: Unabridged Translation of the Third Edition (1744) with the addition of 'Practice of the New Science'.* Cornell University Press, Ithaca, NY.

11. Frankcom, G. & Musgrave, J. H. 1976. *The Irish Giant.* Gerald Duckworth, London, p. 9.

12. Wood, E. J. 1868. *Giants and Dwarfs.* Richard Bentley, London, p. 11.

13. Thorpe, L. (transl. 1966). *The History of the Kings of Britain.* Penguin, London, pp. 52–54.

14. Hadley, C. 2019. *Hollow Places: An Unusual History of Land and Legend.* William Collins, London.

15. Wood, E. J. 1868. *Giants and Dwarfs.* Richard Bentley, London, pp. 2, 11.

16. Haze, X. 2018. *Ancient Giants.* Bear & Company, Rochester, VT, p. 98.

17. See also Bubandt, N. 2014. *The Empty Seashell: Witchcraft and Doubt on an Indonesian Island.* Cornell University Press, Ithaca, NY, pp. 64–65.

18. Forth, G. 1998. *Beneath the Volcano: Religion, Cosmology and Spirit Classification Among the Nage of Eastern Indonesia.* KITLV Press, Leiden, pp. 234–235.

19. Wood, E. J. 1868. *Giants and Dwarfs.* Richard Bentley, London, p. 20.

20. von Franz, M. L. 1995. *Shadow and Evil in Fairy Tales,* revised edition. Shambhala, Boulder, CO, p. 246.

21. Jung, C. G. 2008. *Children's Dreams: Notes from the Seminar Given in 1936–1940.* Princeton University Press, Princeton, NJ, p. 139.

22. Bernheimer, R. 1952 *Wild Men in the Middle Ages.* Harvard University Press, Cambridge, MA, pp. 2–4.

23. Schama, S. 1995. *Landscape and Memory.* Fontana Press, London, p. 97.

24. Simon, E. 2017. *Anthropology of Consciousness* 28: 117–120.

25. Harlan-Haughey, S. 2016. *The Ecology of the English Outlaw in Medieval Literature: From Fen to Greenwood.* Routledge, Abingdon, p. 1.

26. Harrison, R. P. 1993. *Forests: The Shadow of Civilization.* University of Chicago Press, Chicago, IL, p. 80.

27. Borges, J. L. (transl. 1969). *The Book of Imaginary Beings.* E. P. Dutton and Co, New York, NY, preface.

28. MacCulloch, J. A. 1921. *Folk-Lore* 32: 228.

29. Hoskins, J. 1998. *Biographical Objects: How Things Tell the Stories of People's Lives.* Routledge, New York, NY and London, p. 70.

30. Van Gulik, R. H. 1967. *The Gibbon in China: An Essay in Chinese Animal Lore*. EJ Brill, Leiden.

31. Forth, G. 2008. *Images of the Wildman in Southeast Asia: An Anthropological Perspective*. Routledge, Abingdon, pp. 18, 40.

32. Forth, G. 1998. *Beneath the Volcano: Religion, Cosmology and Spirit Classification Among the Nage of Eastern Indonesia*. KITLV Press, Leiden, p. 234.

33. Ting, N. 1978. *A Type Index of Chinese Folktales*. Suomalainen Tiedeakatemia, Helsinki.

34. Campbell, J. 1988. *The Hero with a Thousand Faces*. Paladin, New York, NY, p. 79.

35. Bernheimer, R. 1952. *Wild Men in the Middle Ages*. Harvard University Press, Cambridge, MA, p. 33.

36. Bower, U. G. 1950. *Naga Path*. John Murray, London, p. 128.

37. Bernheimer, R. 1952. *Wild Men in the Middle Ages*. Harvard University Press, Cambridge, MA, p. 44.

38. Brontë, C. (1985, ed. Q. D. Leavis). *Jane Eyre*. Penguin, London, p. 53.

39. Ibid., p. 154.

40. Ibid., p. 483.

41. Aubrey, J. 1881. *Remains of Gentilisme and Judaisme, 1686–87*. W. Satchell, Peyton & Co, London, pp. 67–68.

42. Wavell, S. *et al.* 1966. *Trances*. George Allen & Unwin, London, p. 21.

43. M. MacNeill, cited in B. G. Rieti, 1990. *Newfoundland Fairy Traditions: A Study in Narrative and Belief*. PhD thesis, Memorial University of Newfoundland, p. 40.

44. Woods, H. R. 2022. *Rule, Nostalgia: A Backwards History of Britain*. W. H. Allen, London.

45. Yeats, W. B. 1981. *The Celtic Twilight*. Colin Smythe, Gerrards Cross, p. 158.

46. Conan Doyle, A. 1922. *The Coming of the Fairies*. Hodder & Stoughton, London, pp. 134–135.

47. Hoskins, J. 1993. *The Play of Time: Kodi Perspectives on Calendars, History, and Exchange*. University of California Press, Berkeley and Los Angeles, CA, p. 118.

48. Silver, C. 1986. *Browning Institute Studies* 14: 149.

49. Grinsell, L. V. 1976. *Folklore of Prehistoric Sites in Britain*. David & Charles, Newton Abbot, p. 20.

50. Bernheimer, R. 1952. *Wild Men in the Middle Ages*. Harvard University Press, Cambridge, MA, p. 20.

51. Green, R. F. 2016. *Elf Queens and Holy Friars: Fairy Beliefs and the Medieval Church*. University of Pennsylvania Press, Philadelphia, PA, p. 2.

52. Latham, M. 2020. *The Bookseller's Tale*. Penguin, London, p. 234.

53. Bernheimer, R. 1952. *Wild Men in the Middle Ages*. Harvard University Press, Cambridge, MA, pp. 50, 74, 84, 100.

54. Messner, R. (transl. 2000). *My Quest for the Yeti: Confronting the Himalayas' Deepest Mystery*. Macmillan, London, p. 78.

55. Hadley, C. 2019. *Hollow Places: An Unusual History of Land and Legend*. William Collins, London, p. 326.

56. Engelke, M. 2017. *Think Like an Anthropologist*. Pelican, London, p. 273.

57. G. Forth. 2008. *Images of the Wildman in Southeast Asia: An Anthropological Perspective*, Routledge, Abingdon, p. 97.

Chapter 11: The Perfect Island – A Fairy Tale for Biologists

1. Morris, E. 2018. *The Ashtray (Or The Man Who Denied Reality)*. University of Chicago Press, Chicago, IL, p. 82.

2. Pisani, E. 2014. *Indonesia Etc.: Exploring The Improbable Nation*. Granta, London, pp. 55–56.

3. Messner, R. (transl. 2000). *My Quest for the Yeti: Confronting the Himalayas' Deepest Mystery*. Macmillan, London, p. 156.

4. Ibid.

5. Seabrook, W. 1929. *The Magic Island*. Harcourt, Brace and Company, New York, NY, p. 92.

6. Thomas, K. 1971. *Religion and the Decline of Magic*. Weidenfeld & Nicolson, London.

7. Price, H. & Lambert, R. S. 1936. *The Haunting of Cashen's Gap*. Methuen & Co, London.

8. Attributed to Frank Muir.

9. Latour, B. (transl. 1993). *We Have Never Been Modern*. Harvard University Press, Cambridge, MA.

10. This idea is explored further in W. Stoczkowski, 2002. *Explaining Human Origins: Myth, Imagination and Conjecture*. Cambridge University Press, Cambridge.

11. Wells, C. 1964. *Bones, Bodies and Disease*. Thames & Hudson, London, p. 21.

12. Hoskins, J. 1993. *The Play of Time: Kodi Perspectives on Calendars, History, and Exchange*. University of California Press, Berkeley and Los Angeles, CA, p. 8.

13. Wood Jones, F. 1910. *Coral and Atolls*. Lovell Reeve & Co, London, p. 1.

14. Loh, J. & Harmon, D. 2014. *Biocultural Diversity: Threatened Species, Endangered Languages*. WWF Netherlands, Zeist.

15. Abbi, A. 2020. The pandemic also threatens endangered languages, blogs .scientificamerican.com/voices/the-pandemic-also-threatens-endangered-languages/

Index

Aarne, Antti   164
Abel, Othenio   134
Aboriginal Australians   186, 188,
     189, 206–7, 221, 223, 224
amblypygids   99
Amblyrhiza   103
ana ula people   162–3
Andaman Islands/Andamanese   60,
     62, 264
Andros Island, Bahamas   206
animal burial rituals   169–70
Annius of Viterbo   242
Anselm of Bec   13–15
anthracotheres   45
Aristotelian philosophy   128
Aristotle   203, 209–10
artistic expression, evolution
     of   64–5
Aru Islands   39, 40
Ascension Islands   30–1
Aubrey, John   229, 248
Australia   39, 41, 71, 102, 103, 186,
     188, 189
     see also Aboriginal Australians
australopithecines   69
Austronesian people   61, 159–61,
     186, 190–1, 226–7, 230
aye-aye   27
Aztecs   136, 152, 153–4

Bacon, Francis   97
Bai Juyi   235
Balearic Islands   28, 30
Bali   40, 41–2
Bali and Lombok faunal
     boundary   41–2
Banks, Joseph   203
Barnes, Paul   192, 230, 231

bats   48–9, 102, 173, 186–7
Baucau people   160
Bauer, Aaron   205–6
Bayanov, Dmitri   89
beavers   179, 213
Becanus, Johannes Goropius   136
Beeckman, Daniel   117
Begum the rhino   23
Bemmelen, Rein van   38
Berger, Lee   59
Bergh, Gert van den   191–2
betel nut/*sirih pinang*   11, 13, 80–2,
     84, 141, 142, 145, 195
the Bible and Christianity   126–7,
     128–9, 152, 241–3, 250
biodiversity and traditional
     cultures   203–15, 264–5
birds
     Ascension Islands   30–1
     Bali and Lombok   41–2
     Bornean bristlehead   38
     cave swiftlets   99, 100
     dodo   115, 213
     flightless rails   30–1, 34
     gadfly petrel   105, 107
     giant marabou stork, Flores   46
     gyrfalcon   181
     Haast's eagle   206
     impact on population by
          rodents   104–5, 107
     island nutrient cycling   107
     loss of flight   28, 30–1
     moa   28, 207–8
     native hen   186
     New Zealand snipe   207
     owls and fossil deposits   109–11,
          181
     Scottish crossbill   24–5

slow-growing   28
Stephens Island wren   33
on Sumba   110–11, 113, 146, 172
Tristan albatross   104, 107
Boccaccio, Giovanni   134
bog bodies   241
Bontius, Jacobus   117, 118
Borges, Jorge Luis   209, 245
Borneo   37, 38, 52, 114, 158
Bourtsev, Igor   89
Braad, Christopher   152
Briggs, Katharine   249
British Isles biodiversity   21, 24–5
Brontë, Charlotte   248
Brown, Peter   69, 70
Buckland, William   51, 137
Bulmer, Susan   48
Bulmer's fruit bat   48–9
Burian, Zdeněk   64
Burnett, James, Lord
    Monboddo   117, 149, 152, 243
Burney, David   83
burning as land management   187–9,
    266
Buru   151

Caesar, Edward VII's pet dog   32
Californian islands   30, 33, 102
Calmet, Reverend Agostino   136
Campbell Island   105
carbon dating   180–1
Cargill, David   204
carnivores   26, 28
Caribbean islands   30, 31, 32, 33–4,
    57–8, 103, 107–8, 111, 151, 177,
    179
cats, feral   33
cave art, prehistoric   71, 266–7
Chagos Archipelago   107
Chambers, Raymond   216
Chapman Andrews, Roy   132
Chatham Islands   164
Chaucer, Geoffrey   248
Chesterton, G.K.   245

China   31, 88, 90, 135, 137, 152–3,
    203, 209, 210, 211–12, 214, 246
'Chinese steamroller' population
    expansion, Neolithic   61, 68,
    114, 158–9, 161, 190–1
Chinese 'wild people'   152–3
Christmas Island   108–9
climate change   183, 185
Cocos Islands   261
Colenso, William   204–5
collagen and subfossil bones   180
Columbus, Christopher   57–8, 107
Conan Doyle, Arthur   249
conquistadors, Spanish   152, 153
conservation biology and the fossil
    record   47–9
continental ecosystems   25, 26
continental islands   20–1, 24
convergence, evolutionary   26–7
Cook, Captain   153, 154, 188–9
Cooke, Siobhán   107
Coon, Carleton   88
coral reefs   107
Cortés, Hernán   153–4
Coryphomys   176
Cotter, Patrick   150
Cottingley Fairy photos   249
Crater Lake origin folklore   223,
    224
'creative explosion,' early modern
    human   64–5
Cree people   134
Crees, Jen   43, 44, 52, 98, 110, 112,
    123, 124
Cririe, James   156–7
Cro-Magnon skeletons   138
Crowley, Brooke   107
cryptozoology   87–92, 157, 204
cultural diversity, modern
    human   262–4
cultural sameness, human   262–3
Current Anthropology   89
Cuvier, Baron Georges   126, 127
cyclopes legend   134, 163, 244

da Vinci, Leonardo   128
Dammerman, Karel   50, 92, 101,
    118, 119
Dampier, William   188
Darwin, Charles   25, 29, 32, 41, 46,
    49, 63
deer mice   33
defence mechanisms, loss of   28
Defoe, Daniel   37
Denisovans   71, 137, 182
Diamond, Jared   61, 106, 159
diet, human   62–3
Dikötter, Frank   153
dinosaurs   130, 131, 135
    dwarfed   30
dodo   115, 213
Doggerland   21
dragons   135, 245
Dubois, Eugène   63, 70
duck-billed platypus   39
dugong   196, 213, 266
Dun Cow legend   169
dwarfing, island   29–30, 58–62, 93
    buffalos   42, 71
    elephants   30, 66, 134, 163, 261
    emus   203
    hippos   29, 83, 137
    proboscideans   29–30, 45

Earl, George Windsor   37, 39, 40
early modern humans   61, 64, 65,
    71–2
Easter Island   105–6, 159
ebu gogo   73–4, 88, 91, 120, 246
echidna, long-beaked   40, 43
    short-beaked   39
Edward VII, King   32
Efé and Mbuti people   60, 61
elephants   129–30, 135, 137, 138
    tiny/dwarf   30, 66, 134, 163, 261
Eliade, Mircea   250
Eliot, T.S.   259
epistemological pessimism vs
    optimism   260–1
Estanque Island   33

ethnotaxonomy   204–15, 265
'Eua   24
Euhemerus   218
Eurasia   45
evolutionary processes   20, 24, 25, 41
    adaptive growth patterns   28–9
    convergence   26–7, 103
    Foster's rule/island law   28–30,
        58–62, 93, 103
    human   58–65
    loss of defence mechanisms   28, 31
    reproductive strategies   28
extinction   31–4, 38, 71–2
    academic acceptance of
        process   126–7
    animals identified in Indigenous
        stories and folklore   205–15
    'extinction amnesia'   214–15
    'extinction filter'   177
    Holocene   185–6
    human agency   31–5, 182–7, 203,
        260, 261
    incorrectly assumed   48–50
    'overkill hypothesis'   183
    reconstructing extinction
        chronologies   179–82
    Signor-Lipps effect/Romeo
        Error   72, 91, 181

fairies and little people   132, 133,
    156–7, 218, 224–5, 247, 248–50
false killer whale/'thick-toothed
    grampus'   49
Fang Yizhi   201
feral children   148–9, 163
Fiji   26, 29, 58, 170, 204
fishing   65
Flannery, Tim   48, 178–9
Flaubert, Gustave   123
Fleck, David   209
Fleure, Professor Herbert   156
Flores   45, 46, 48, 50, 62, 67, 68–70,
    71–4, 81, 94, 104, 110, 114, 120,
    139, 161, 162–3, 165, 173, 176, 177,
    182, 190, 192, 204, 243, 261–2

Flowers Barrow, Dorset 240
folklore
  'disenchantment' 259
  and fact 215–26, 257–8
  modern Western culture 259
  repetitive themes in global 240–50
  see also ebu gogo; fairies and
    little people; giants, real and
    mythological; mili mongga;
    wildmen/man-beasts
Forster, E.M. 244
Forth, Gregory 35, 52, 72–5, 87, 94,
  118–19, 120, 139, 151, 204, 213,
  252
fossils, alternative explanations
    for 133–4, 260
  historical explanations for 127–30
  practical and artistic use of 130–2
  as religious/sacred relics 137–8
  as remains of monsters and fantastical
    beasts 134–7
  role in conservation biology 47–9
  spiritual and magical beliefs
    about 132–3
Foster's rule/island law 28–9, 45,
  58–62, 93, 103
Fowler, Cynthia 186–7
Frankcom, G. 150
Frazer, Sir James George 145
frogs, endemic Fijian 26
fruit bats 48–9, 186–7

Galápagos Islands 29, 30
Gaunilo 14, 15
Gee, Henry 74
Geoffrey of Monmouth 221–2,
  243
geological events in oral tradition/
  geomythology 222–5
Gerald of Wales 241
giant lemurs 83
giant marabou stork 46
Giant of Castelnau, France 240
giants, real and mythological 137–8,
  150, 155, 240–4

giant rats 28, 45, 66, 101, 103,
  109–10, 113, 176–9, 190
  new species found on
    Sumba - *Raksasamys* and
    Milimonggamys 175–6, 180–2,
    190–1, 230, 260, 265
  in Sumbanese folklore 226–7
giant reptile 30, 31–2, 46, 48, 192,
  261
giant stick insect/land lobster 105
giant tortoise 30, 31–2, 46, 48
gibar bohot 151
gibbon, Hainan 214–15
Gigantomachy 206
Gigantopithecus 88–9, 91
Glastonbury Abbey 241
globalisation and cultural
  diversity 263–5
God, existence of 13–15
Godbey, A.H. 195
Gondwana 21–2, 23, 26, 45
gorilla 116
Gough Island 104, 107
Grayson, Donald 184
Great Chain of Being 126
Greece/Greeks, Ancient 134–5, 163,
  218, 222, 246
greenhouse gases 185
Grotte du Trilobite,
  Arcy-sur-Cure 131

Hadley, Christopher 250
Hadrian, Roman Emperor 138
Haeckel, Ernst 63
Hain, János Paterson 135
Hainan 31
Haiti 210–11
Hall, Joseph 216
Halmahera 155
Hamilton, Alexander 117
Hanno 116
Hansford, James 43, 82, 98, 101,
  112, 124, 173
Hartlepool 118
Haţeg Basin, Romania 30

Hauser, Kaspar   163
Hawaiian archipelago   22, 58, 106, 154
headhunting   171
Heavenly Emporium of Benevolent Knowledge (J.L. Borges)   209
Henige, David   219, 224
Henrion   242–3
Hensbroek, Jacques van Steyn van   204
Hermetic philosophy   128
Hesse, Hermann   11
Heuvelmans, Bernard   88–9, 157
Hillary, Sir Edmund   93
Hindustan, SS   108–9
hippos   29, 83, 137
Hispaniola   107–8, 179, 210–11
Historia Animalium (Aristotle)   203
Holocene Epoch   21, 184–6
hominins, early   63–4, 65–7, 68, 71, 137, 182, 261–2
   see also Homo erectus, Homo floresiensis, Homo luzonensis; Neanderthals
Homo erectus   63–4, 65–7, 93
Homo floresiensis   68–70, 71–4, 88, 93, 152, 161, 182, 246, 261, 262
Homo luzonensis   70, 93, 182
Hoskins, Janet   63, 200, 226, 246, 261
hurricanes   23
hutia   32–3, 71, 107–8, 111, 179, 210
Hyorhinomys stuempkei/Sulawesi snouter   103

Ice Age glaciation cycles   21, 37–8, 42, 103, 182, 183–4
iguanas   23
Iliad (Homer)   222
immunity and island communities   56–7
In Behalf of the Fool (Gaunilo)   13
Indian Ocean archipelagos   30
   see also Galápagos Islands
Indigenous stories/folklore and fact   215–26

Indonesia   35, 37, 38–41, 42–3, 55–6, 111–12, 161, 165
Indonesian Throughflow   42, 50
invasive species, destructive   103–9, 115
Iron Age archaeology   130–1
islands
   animal vulnerability/ extinction   31–3
   bird-mediated nutrient cycling   107
   continental islands   20–1, 24–5
   evolutionary processes   20, 24, 25, 26–9, 58–61
   geological time   21
   human cultural diversity   55–6
   human growth/size   58–62, 93, 102
   hybrid formation processes   24
   microcontinents   22, 25, 40
   oceanic   22
   overwater colonisation/ dispersal   22–3, 26, 42, 46, 50, 67, 75–6, 102, 114
   power/attraction of   14, 19–20
   tidal islands   21
Isle of Man   248

Jacquemart Island   105
Jamaica   23
Java   37, 39, 40, 63–4, 65–6, 67, 70, 117, 133
Java Man/*Homo erectus*   63–4, 65–6, 67
Jefferson, Thomas   127, 128, 130
Jeffree, Tim   43, 51, 82, 98, 100, 112, 113, 124, 125, 173, 192, 196, 231, 235
John Frum cult   155
Johnson, Samuel   57
Johnston, Sir Harry   204
Jung, Carl   244

Kachari people   247
Kalevala   222
katoda – skull tree   171

kaweau/kawekaweau   205–6
Keightley, Thomas   248
Kennerley, Ros   179
Kerl, Johann   180
kidoky   83
kilopilopitsofy   83
King Island, Australia   203
King Kong film (1933)   30
Kiöping, Nils Matsson   117–18
Kipling, Rudyard   248
Kircher, Athanasius   128, 135
Kirkdale Cavern, Yorkshire   51, 137
Klamath Native Americans   223,
    224
Koenigswald, Ralph von   64, 65–6,
    88, 131
Komodo dragons   46, 48, 173, 192,
    204, 261
Krakatau/Krakatoa   40
Krantz, Grover   89, 92
Krappe, Alexander Haggerty   162,
    163

landscape burning   187–9, 266
language diversity   263–4
Lartet, Édouard   51
Late Pleistocene Epoch   21, 182–3
Latour, Bruno   259
Lawrence, D.H.   19
Lazarus taxa   72, 91
Leahy, Michael   154–5
Leclerc, Georges-Louis, Comte de
    Buffon   130
Leguat, François   151
lemurs   26–7, 83
Lesser Antilles   102
Lesser Sunda Islands   40, 43, 52, 62,
    160, 188–9
Levi-Strauss, Claude   205, 210, 247
Lhuyd, Edward   128, 213
Liang Bua cave, Flores   68–9, 71,
    110, 197, 261
Linnaeus, Carl   117–18, 151
Lio people   74

Little Swan Island   32–3
living fossils   49–50
Llwyd, Humphrey   179
Lombo/Karendi people   162
Lombok   41, 47
Lombok Strait   42
long-tailed macaque   114, 115–16,
    119, 120
Lord Howe Island   105
Lovecraft, H.P.   163
Lowe, Percy   32–3
Luplupwintem cave, New
    Guinea   49
Lyall, David   33

macaque   114–16, 119, 120
MacCulloch, Canon John   156,
    245–6
Machen, Arthur   218
MacKinnon, John   93
Maclean, Charles   57
Madagascar   22, 26–7, 29, 83, 157,
    206, 208
Makassarese people   165
Malay Archipelago   41, 165
Malaysia   161
Malson, Lucien   149
Maluku/Molucca Islands   40, 41,
    152
mammalian carnivores   26
mammoth fossils   126, 128, 129, 130,
    133, 135, 137, 139
Maniq people   161
Māori people   164, 170, 205–6,
    207–8, 213, 221
Marapu religion   197, 232
Marco Polo   119
Marsh, Othniel Charles   135
marsupials   39, 41, 43
Martyr, Debbie   93
mastodon fossils   130
Matius   11, 12–13, 253–4
Matses Indians   209
Mauritius   115

Mayor, Adrienne 134–5
Mayr, Ernst 205
medicine trade, traditional 35, 88, 131
Mediterranean Islands 29–30
megafauna 182–3, 188, 206, 207
megalithic tombs 85, 97, 140, 147, 163, 170
Megalosaurus 130
Meganthropus palaeojavanicus 65–6
Melanesian Islands 61, 155
Melville, Herman 175
Merlin legend 132
Messner, Reinhold 258
meteor strikes 222, 224
meu rumba 84–5, 94, 121, 168–9, 237
microcontinents 22, 25, 40
Middleton, John 150
mili mongga 53, 75–7, 175–6, 226, 250–1, 257–8
  appearance 75, 79, 83, 199, 237
  bones and graves 85, 94, 121, 124–6, 137, 139–40, 145–7, 168, 200, 201, 202, 232–3, 253–4
  as a cannibal 167, 200, 237
  captured child stories 76, 83
  cave of the mili mongga ('Gua Mili Mongga') 85
  as human deviant/'other' 148
  interpreted as a biblical giant 198
  magical powers and salt 252
  megalithic tomb carvings 85
  'meu rumba' story 84–5, 94, 121, 168–9, 237
  misinterpreted encounters with different ethnic groups 162–3, 166
  stone wall construction 162–3
  stories of historical meetings/ sightings 82–3, 86, 147, 167–8, 199–201, 202–3, 216, 236–7, 238
  stories of recent/existing 76, 79, 237, 238, 247–8

story from the rato/shaman of Migurumba 198–201, 202, 247–8
  Sumbanese warfare 168, 169
Milimonggamys 175–6, 181–2, 190–1
Miller, Gerrit 110
moa 28, 207–8, 213
Molyneux, Thomas 130
Monboddo, Lord, James Burnett 117, 149, 152, 243
monotremes 39–40
Montandon, Georges 119
Morris, Errol 257
Mount Tambora 40, 160
Muldashev, Ernst 243
Mundy, Peter 30–1
Musgrave, J.H. 150
Musket Wars 164
Myotragus 28, 29, 30

Nage people 73–4, 120, 139, 176, 178, 190, 243–4, 246
Naish, Darren 92
Nan Madol 221
Narrangga people 223
Natural History (Pliny the Elder) 116
Nature 64, 74
Neanderthals 65, 70, 71, 91, 150, 157, 182
Needham, Rodney 166
Neolithic archaeology 68, 131, 169, 240
Neolithic farmers 61, 68, 114, 158–9, 161, 190–1, 226
Neoplatonic philosophy 126, 128
New Caledonia 23, 26
New Guinea 27, 39, 41, 43, 48–9, 103, 154–5, 160, 165, 178, 205
New Zealand 21–2, 26–7, 28, 29, 31, 105, 159, 164, 170, 204–6, 207–8
Nicobar Islands 117
Nie Huang 203
nittaewo, Sri Lanka 93
Nunn, Patrick 223

Odyssey (Homer)   134
Oglala Sioux, Native American   135
Orang Asli people   161
orang pendek, Sumatra   92–3, 118
orangutans   34, 117–18
Oryx   93
Osman Hill, Professor William   93,
     118
'otherness', human
   encounters with 'different' ethnic
      groups   151–7, 161, 162
   feral children   148–9, 163
   medical/behavioural
      conditions   148, 150
   physical appearance   149–50
   'sub-human' outlaws   150–1
   western visitors and indigenous
      peoples   153–6
   see also mili mongga; wildman
      myths/man-beast stories
outrigger canoes, invention of   159
overwater colonisation/dispersal
     22–3, 26, 42, 46, 50, 67, 75–6,
     102, 114
Owen, Richard   49

Pacific Islands, north   30
Pacific Ring of Fire   40
Pahvant Ute Native Americans   132
Palaeoloxodon falconeri   30
Palau   59–60
Palawan   37, 38
Papagomys armandvillei   176, 177, 178,
     230
Papua and West Papua   39–40,
     226–7
Papua New Guinea   39, 48, 179,
     263
Paré, Ambroise   136
Pastrana, Julia   148
Patagonia   243
patau patuna   238–9
peat-swamp forests   34
Peron, François   203

Peter the Great   129
'Peter the Wild Boy'   149
Philippines   70, 103, 159, 162–3
Phillpotts, Eden   11
Pigafetta, Antonio   151–2, 164, 243
Pingelap, Micronesia   58
Pirahã people   218
Pisani, Elizabeth   257–8
plants and trees
   conifers   24, 26
   Easter Island palm forests   105–6
   fossilised and petrified wood   131–2,
      227–8
   impact of invasive rodent
      species   105–6
   'island law' and size   29
   land clearance   187–9, 266
   Marapu religion   197, 232
   on Sumba   172, 187–9
Plato   128
Pliny the Elder   116
Plot, Robert   129–30, 138
podocarp trees   24
Poetic Edda   243
Polo, Marco   60
Poma people   163
primates and humans, confusion in
     distinction of   115–20
Proust, Marcel   9
Pu Songling   203
pygmy people   58–9
'pygmy-Pict' theory   156

Qata ethnic group   61, 62, 159–60,
     161–2, 226

Rabelais, François   242
Raglan, Lord   147
Raksasamys tikusbesar   175, 181–2,
     190–1, 230, 260
Rampasasa pygmies   61, 161
rats   102–3
   giant rats   28, 45, 66, 101, 103,
      109–10, 113, 176–9, 190

new species found on
   Sumba - *Raksasamys* and
   Milimonggamys   175–6, 180–2,
   190–1, 230, 260, 265
   in Sumbanese folklore   226–7
   invasive species   103–9, 189–91, 265
Ray, John   130
Redmond, Ian   93
Reid, Nicholas   223
Renaissance and Enlightenment
   thinkers   117, 128, 242
reproductive strategies   28
rhinos   23, 38–9, 214
Rhynchocephalia   22
Roti   67
Rousseau, Jean-Jacques   117
Ruddiman, William   185
Russell, Anthony   205–6

Sacks, Oliver   58
Sahul   103, 158
Sahul Shelf   39, 40, 42
Saint Kilda, Outer Hebrides   57, 109
Saint Michael's Mount,
   Cornwall   240
saki monkey   209
sandalwood trade   164, 189, 191
Sartono, S.   51
sasquatch   88–9, 90, 93, 250
Schama, Simon   245
Scheuchzer, Johann   129
Schliemann, Heinrich   222
Scott, Walter   156
Seabrook, William   259
Shakespeare, William   9
shifting baseline syndrome   212–13
Siberian islands/Siberia   30, 224
Signor-Lipps effect/Romeo
   Error   72, 91
Simons, Elwyn   68
sirih pinang see betel nut
size/growth, 'island law'   28–30, 45,
   58–62, 93, 103
slave trade   165–6

Sloane, Sir Hans   134
sloth lemurs   26–7, 206
sloth, ground   30
Solomon Islands   102
Somerset, Fitzroy Richard, 4th Lord
   Raglan   216
Southeast Asia   34–5, 38, 43, 61, 62,
   68, 117
   early hominins   63, 66
   early modern humans   61–2, 65,
      71
   islands   30, 35, 43, 150, 151–2
   Neolithic farmers   61, 68, 114,
      158–9, 161, 190–1, 226
South Georgia   104
Stegodon sumbaensis   47, 50, 137
stegodons/stegodon fossils   45–6, 47,
   48, 50, 61–2, 67, 69, 132, 137,
   139, 191–2, 261
Steiner, Gerolf   103
Steno, Nicolas   257
Sterckx, Roel   210
Sterndale, Robert   152
Stevenson, Robert Louis   58
Stewart, James   90, 93, 250
Stiernhielm, Georg   129
Stonehenge   163, 221–2
stone tools, prehistoric   67, 133, 230,
   231–2
stone wall construction and
   'otherness'   162–3
Stoney Littleton long barrow,
   Somerset   131
Strabo   131
Sulawesi   40, 42–3, 45, 67, 71, 103,
   114, 141, 158, 165, 266–7
Sumatra   37, 38–9, 41, 52, 92–3, 118,
   119
Sumba   12–13, 43–4, 265–7
   19th century smallpox epidemic   58
   Austronesian people   61, 158–9,
      160–1, 190–1, 226–7, 230
   burial/funeral practices   141–2, 145
   climate   52, 187

cultural diversity 56

discovery of human remains 235, 239–40

exposure to different ethnic groups/ colonisation 158–9, 160–1, 162

folklore/oral tradition (other than mili mongga) 197, 220, 226, 238–9, 252–3, 258–9

fossil wood beds 227–8

geology 50, 51

giant rat extinction 186–7, 190, 230, 260, 265

giant rat folklore 226–7

giant rat fossils – *Raksasamys* and *Milimonggamys* 101, 109–10, 113, 175–6, 180–2, 190–1, 230, 260

Goat Cave 114, 115

headhunting 171

human introduction of animal species 114–15

hunting practices 186–7, 189–90

Karel Dammerman's records of fauna 50

Komodo dragon fossils 191–2, 261

Lambanapu village 239–40

landscape burning 189–90, 266

Liang Kanabu Wulang/Falling Moon Cave 172–3

Liang Lawuala cave 98–101, 109–10, 175, 181, 186–7, 190–1

Lombo/Karendi people 226

macaque population 114–15, 120

Mahaniwa village 98, 111–13, 121

Mangili village 146, 170

Marapu religion 197, 232, 266

megalithic tombs 85, 97, 170

Migurumba village 198–202

Miurumba village 84–5

nyale worm harvest 62–3

owls – fossils and fossil deposits 109–10, 181

Pasola festival 44, 201

patau patuna 238–9

prehistoric stone tools 230, 231–2

Qata population 162

rato/shamen of Migurumba 198–202

Rindi village 140–1, 147

ritual animal burials 170

sandalwood trade 164, 189, 265

slave trade 165–6

stegodon fossils 47, 50, 137, 191–2, 261

Varanus hooijeri fossils 110, 261

Verhoeven's treasure map 197–8, 229

Waikabubak village 170–2, 228–9

Waingapu 44, 123, 166–70, 196

wall construction 162, 164

Watumbaka village 47, 51–2

wedding party 231–3

Wunga village 226–7

Sumbawa 40, 47, 118

Sunda Shelf 37–8, 40, 41, 42, 63–4, 65–6, 114

Sunda Shelf and Sahul Shelf, islands between 41, 42–3
    see also Flores; Sulawesi; Sumba; Timor; Wallacea

Sundaland 38, 66

Sykes, Bryan 149–50, 157

Taino people 57–8, 109

Taiwan 161

Tanimbar 216–17

Tasmania 186

tectonic forces 21, 22, 25

Thompson, Christina 159, 220

Thompson, Stith 164

Throwim Way Leg (T. Flannery) 48–9, 178–9

thylacine 39, 186

tidal islands 21

Timor 43, 45, 46, 48, 65, 67, 160, 176–7, 188–9, 191

Toba supervolcano 41

Tongan archipelago 24

Torajan people 141